Edexcel GCSE
Mathematics B
Higher
Student Book
Unit 1

Series Director: Keith Pledger
Series Editor: Graham Cumming

Authors:
Chris Baston
Julie Bolter
Gareth Cole
Gill Dyer
Michael Flowers
Karen Hughes
Peter Jolly
Joan Knott
Jean Linsky
Graham Newman
Rob Pepper
Joe Petran
Keith Pledger
Rob Summerson
Kevin Tanner
Brian Western

ROSEBERY   SCHOOL
EPSOM

DATE		FORM

A PEARSON COMPANY

Published by Pearson Education Limited, a company incorporated in England and Wales, having its registered office at Edinburgh Gate, Harlow, Essex, CM20 2JE. Registered company number: 872828

Edexcel is a registered trademark of Edexcel Limited

Text © Pearson Education Limited 2010

The rights of Chris Baston, Julie Bolter, Gareth Cole, Gill Dyer, Michael Flowers, Karen Hughes, Peter Jolly, Joan Knott, Jean Linsky, Graham Newman, Rob Pepper, Joe Petran, Keith Pledger, Rob Summerson, Kevin Tanner and Brian Western to be identified as the authors of this Work have been asserted by them in accordance with the Copyright, Designs and Patent Act, 1988.

First published 2010

13 12 11 10
10 9 8 7 6 5 4 3 2

British Library Cataloguing in Publication Data
A catalogue record for this book is available from the British Library.

ISBN 978 1 846900 91 4

Typeset by Techset, Gateshead
Picture research by Rebecca Sodergren

Acknowledgements
The publisher would like to thank the following for their kind permission to reproduce their photographs:
Alamy Images: Martin Phelps 29; **Corbis**: Régis Bossu 117, LWA-Sharie Kennedy 93, Mike McGill 1; **iStockphoto**: Floyd Anderson 141; **Photolibrary.com**: Joff Lee 60, Kathy deWitt 113; **shutterstock**: 113 (2), Kuttly 112; **Wales on View**: 112 (2), 112–113

All other images © Pearson Education

We are grateful to the following for permission to reproduce copyright material:

Tables
Table in Exercise 19.I adapted from 'CO2 emissions 2003-2008', Crown Copyright material is reproduced with the permission of the Controller, Office of Public Sector Information (OPSI).; Table in Exercise 19.I adapted from 'Agricultural Survey of Cattle, England 2006-2008', Crown Copyright material is reproduced with the permission of the Controller, Office of Public Sector Information (OPSI).

Every effort has been made to trace the copyright holders and we apologise in advance for any unintentional omissions. We would be pleased to insert the appropriate acknowledgement in any subsequent edition of this publication.

Disclaimer
This material has been published on behalf of Edexcel and offers high-quality support for the delivery of Edexcel qualifications.
This does not mean that the material is essential to achieve any Edexcel qualification, nor does it mean that it is the only suitable material available to support any Edexcel qualification. Edexcel material will not be used verbatim in setting any Edexcel examination or assessment. Any resource lists produced by Edexcel shall include this and other appropriate resources.

Copies of official specifications for all Edexcel qualifications may be found on the Edexcel website: www.edexcel.com

Contents

About this book

All set to make the grade!

Edexcel GCSE Mathematics is specially written to help you get your best grade in the exams.

> Recap with a skills check at the start of a section – make sure you're up to speed.

> Section objectives show what you'll be learning.

> Loads of practice to help you feel secure before you move on.

> Graded questions – so you know what you're achieving.

> Full coverage of the new-style assessment objective questions – AO2 and AO3.

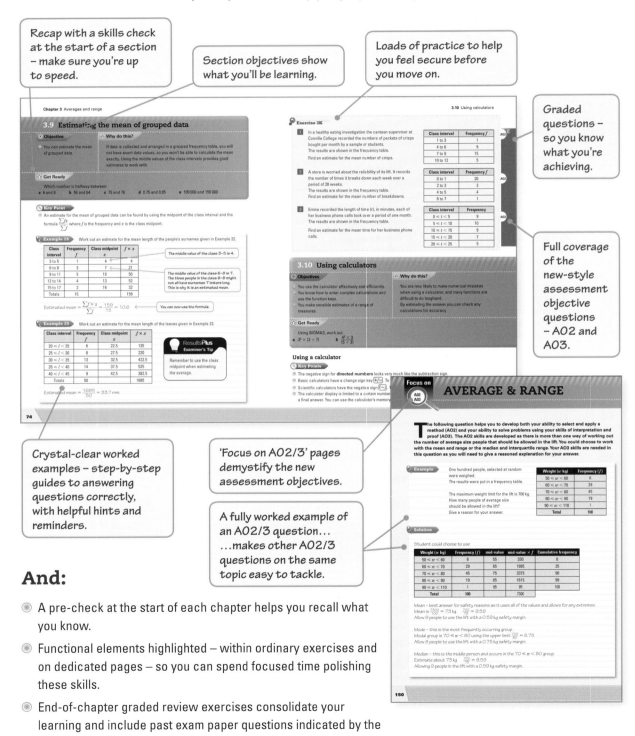

> Crystal-clear worked examples – step-by-step guides to answering questions correctly, with helpful hints and reminders.

> 'Focus on AO2/3' pages demystify the new assessment objectives.

> A fully worked example of an AO2/3 question... ...makes other AO2/3 questions on the same topic easy to tackle.

And:

- ◉ A pre-check at the start of each chapter helps you recall what you know.
- ◉ Functional elements highlighted – within ordinary exercises and on dedicated pages – so you can spend focused time polishing these skills.
- ◉ End-of-chapter graded review exercises consolidate your learning and include past exam paper questions indicated by the month and year.

About ActiveTeach

Use **ActiveTeach** to view and present the course on screen with exciting interactive content.

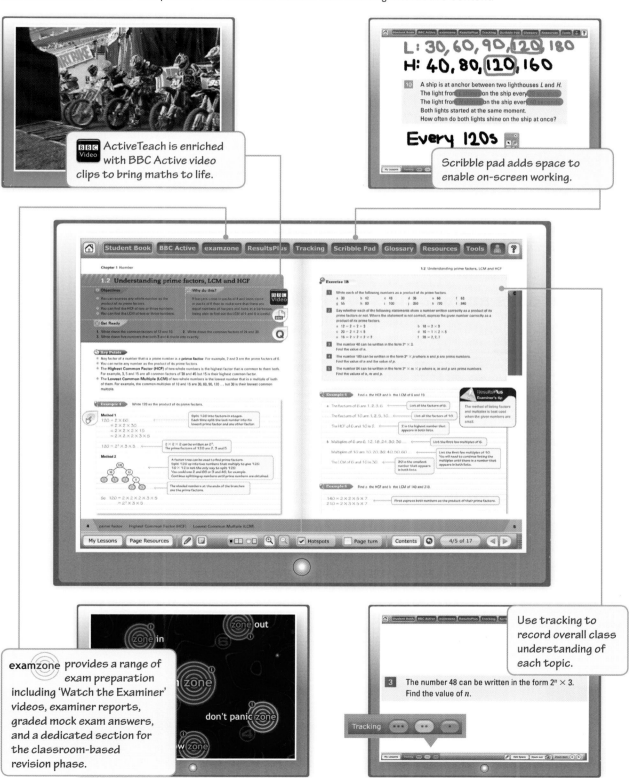

ActiveTeach is enriched with BBC Active video clips to bring maths to life.

Scribble pad adds space to enable on-screen working.

examzone provides a range of exam preparation including 'Watch the Examiner' videos, examiner reports, graded mock exam answers, and a dedicated section for the classroom-based revision phase.

Use tracking to record overall class understanding of each topic.

About Assessment Objectives

Assessment Objectives define the types of question that are set in the exam.

Assessment Objective	What it is	What this means	Range % of marks in the exam
A01	**Recall** and use knowledge of the prescribed content.	Standard questions testing your knowledge of each topic.	45-55
A02	**Select** and apply mathematical methods in a range of contexts.	Deciding what method you need to use to get to the correct solution to a contextualised problem.	25-35
A03	**Interpret** and analyse problems and generate strategies to solve them.	Solving problems by deciding how and explaining why.	15-25

The proportion of marks available in the exam varies with each Assessment Objective. Don't miss out, make sure you know how to do AO2 and AO3 questions!

What does an AO2 question look like?

D **AO2**

16 Katie wants to buy a car.
She decides to borrow £3500 from her father. She adds interest of 3.5% to the loan and this total is the amount she must repay her father. How much will Katie pay back to her father in total?

> This just needs you to
> (a) read and understand the question and
> (b) decide how to get the correct answer.

What does an AO3 question look like?

D **AO3**

17 Rashida wishes to invest £2000 in a building society account for one year. The Internet offers two suggestions. Which of these two investments gives Rashida the greatest return?

> Here you need to read and analyse the question. Then use your mathematical knowledge to solve this problem.

CHESTMAN BUILDING SOCIETY
£3.50 per month
Plus **1% bonus** at the end of the year

DUNSTAN BUILDING SOCIETY
4% per annum. Paid yearly by cheque

Focus on

A02 A03

We give you extra help with AO2 and AO3 on pages 148–153.

About functional elements

What does a question with functional maths look like?

Functional maths is about being able to apply maths in everyday, real-life situations.

GCSE Tier	Range % of marks in the exam
Foundation	30-40
Higher	20-30

The proportion of functional maths marks in the GCSE exam depends on which tier you are taking. Don't miss out, make sure you know how to do functional maths questions!

In the exercises…

20 The Wildlife Trust are doing a survey into the number of field mice on a farm of size 240 acres. They look at one field of size 6 acres. In this field they count 35 field mice.

a Estimate how many field mice there are on the whole farm.

b Why might this be an unreliable estimate?

> You need to read and understand the question. Follow your plan.
>
> Think what maths you need and plan the order in which you'll work.
>
> Check your calculations and make a comment if required.

...and on our special functional maths pages: 154–157!

Quality of written communication

There will be marks in the exam for showing your working 'properly' and explaining clearly. In the exam paper, such questions will be marked with a star (*). You need to:

- use the correct mathematical notation and vocabulary, to show that you can communicate effectively
- organise the relevant information logically.

ResultsPlus

ResultsPlus features combine exam performance data with examiner insight to give you more information on how to succeed. ResultsPlus tips in the **student books** show students how to avoid errors in solutions to questions.

Watch Out!

Some students use the term average – make sure you specify mean, mode or median.

This warns you about common mistakes and misconceptions that examiners frequently see students make.

ResultsPlus Exam Question Report

91% of students scored poorly on this question because they did not use the midpoint of the range to find the mean of grouped data.

This gives a breakdown of how students did on real past exam questions.

ResultsPlus Examiner's Tip

Make sure the angles add up to 360°.

This gives exam advice, useful checks, and methods to remember key facts.

ResultsPlus in the **ActiveTeach** provides interactive practice for AO2 and AO3 questions...

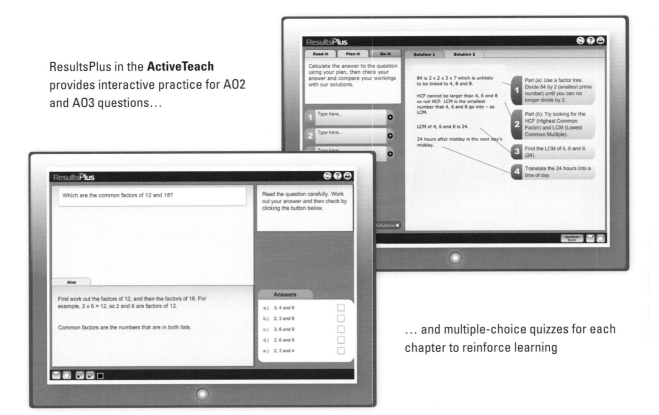

... and multiple-choice quizzes for each chapter to reinforce learning

1 COLLECTING AND RECORDING DATA

A local council wants to know whether the facilities for teenagers are adequate in the town.

How could it find out people's views?

How could these views be recorded and presented?

When you have read this chapter you will know how this can be done.

◎ Objectives

In this chapter you will:

- learn about the statistical problem-solving process and consider different types of data
- discover how to collect, record and interpret data
- look at various sampling methods
- learn how to identify possible sources of bias.

1.1 Introduction to statistics

Objectives

- You can understand the stages of an investigation.
- You can formulate a question in terms of the data needed.
- You can classify data as qualitative (categorical) or quantitative (numerical).
- You can classify quantitative data as discrete or continuous.
- You can choose appropriate units of measurement and convert between metric units.

Why do this?

To find out how good teachers are at predicting the grades their students will get in an exam, you could carry out a statistical investigation.

Get Ready

How can you find the following information?

a The average amount of lunch money for your classmates.

b What flights there are from Manchester to Washington D.C.

c How many people voted for the Green Party in the last election.

Classifying data

Key Points

- **Statistics** is used to provide information. The statistical problem-solving process can be shown as a simple diagram:

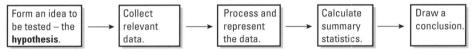

- Data that you collect yourself is called **primary data**; data collected by other people is called **secondary data**.
- **Qualitative data** can be described in words. For example, the colours of shirts on sale in a shop.
- **Quantitative data** are numerical observations. There are two types:
  - **Discrete data** can only take certain numerical values. For example the number of carriages on trains.
  - **Continuous data** can take any numerical value. For example weights, times, lengths and temperatures are continuous.

Metric units

- Continuous data is collected in **metric units**.
- The units of measurement used in the UK are metric units.

Measurement	Length	Area	Weight	Capacity/Volume
Basic unit	metre (m)	square metre (m²)	gram (g)	litre (*l*)

- The word for each basic unit can be changed into a bigger or smaller unit by adding one of the following words to the front of it.

Word	milli	centi		kilo
Meaning	$\frac{1}{1000}$	$\frac{1}{100}$	1	1000
Example	millimetre $= \frac{1}{1000}$ metre	centimetre $= \frac{1}{100}$ metre	metre	kilometre $= 1000$ metres

- To change between metric units you only need to multiply or divide by 10, 100 or 1000.
- To change from smaller units to larger ones you divide.

⦿ To change from larger units to smaller units you multiply.
For example, to convert lengths

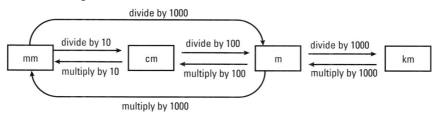

⦿ When measuring an object, select an appropriate unit of measure. For example, use metres for the height of a house, kilograms for the weight of a dog and litres for the amount of petrol in a tank.

Example 1

An estate agent collects the following information about houses for sale.

Type of house	Number of bedrooms	Garden area	Price
Detached	4	390 m²	£321 000
Semi-detached	3	170 m²	£184 000
Terraced	3	150 m²	£177 000
Flat	2	0	£196 000

Describe the data in each column as qualitative or quantitative. If quantitative, state whether it is discrete or continuous.

Type of house: qualitative
Number of bedrooms: quantitative and discrete
Garden area: quantitative and continuous
Price: quantitative and continuous

> See if the data item can be represented as a number. If it cannot it is qualitative data. If it can be given as a number it is quantitative. If quantitative ask yourself 'Can it only take certain values?' If it can it is discrete; otherwise it is continuous.

Exercise 1A

Questions in this chapter are targeted at the grades indicated.

1 Write down whether each of the following is secondary or primary data.
  a Data collected by you from a government website
  b Data collected by you from a newspaper
  c Data collected by you questioning people in a shopping centre

2 Write down whether the following are qualitative or quantitative data.
  a The numbers of students in classes
  b The colour of students' eyes
  c The weight of dogs
  d The floor area of houses

3 Write down whether the following are continuous or discrete data.
  a The number of trees in a wood
  b The time taken to run 100 m
  c The length of flower stems
  d The number of animals in a zoo

4 Kai knows that he needs 2.6 metres of material to make one curtain.
Work out how many metres he needs to make 10 curtains.

5 A store owner orders a roll of material which is 92 metres long.
It is to be cut into 100 equal size pieces. How long will each piece be?

6   A garage forecourt fuel tank holds 56 000 litres of petrol.
    The fuel tank of a certain make of lorry holds 200 litres of petrol.
    What is the maximum number of empty lorries that can be filled from the garage forecourt fuel tank?

7   Mr Longton buys 456 metres of flex. At a later date he buys 518 metres of flex.
    He wants as many pieces of flex that are 10 metres long as possible.
    How many 10 metre lengths of flex can he make?

8   Mira is going to make school skirts for her three daughters.
    The table shows the length of material, in centimetres, that she needs for each skirt.

Child	Latika	Nirupa	Saria
Length of material (cm)	130	156	183

    a Work out how many centimetres of material Mira needs altogether.
    b Work out how many metres of material Mira needs altogether.

    Mira buys 8 metres of material.
    c Work out whether or not she has enough material to make one or more extra skirts for her daughters.

9   A medical researcher wants to find out how effective Drug A is at curing malaria.
    a Write down a hypothesis he could use.
    b What is the next thing that he would need to do?

10 It takes 125 g of flour to make a sponge cake.
    Deirdre wants to make 15 sponge cakes for the village fete.
    Maria has a 1.5 kg bag of flour. Investigate whether she has enough flour to make the 15 cakes.

11 Lemonade is supplied in 2 litre bottles.
    A lemonade glass holds 80 ml.
    a Work out how many glasses can be filled from a 2 litre bottle.

    A bottle of lemonade costs £1.65 from a supermarket.
    b How much would it cost to fill 80 glasses?

1.2 Sampling methods

⊙ Objectives

- You can collect information about a population by using a sample.
- You can select a simple random sample.
- You know that in a simple random sample each member of the population has an equal chance of being selected.

⊘ Why do this?

A city council wants to know how many people are likely to support the idea of building a swimming pool. They can't ask everybody in the city but they can ask a sample of people.

⬦ Get Ready

If it takes 15 seconds for one student to answer a question, how long would it take to get answers from everyone in your class?

Key Points

◉ A small, but carefully chosen, number of people can be used to represent the **population** of a country. These chosen individuals are called a **sample** and the investigation itself is called a sample survey.

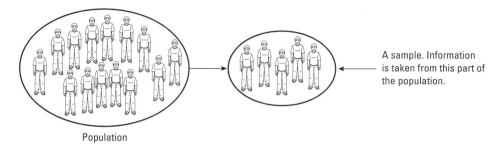

A sample. Information is taken from this part of the population.

Population

◉ The sample must be **representative** of all the people or items being investigated, with each member of the population having an equal chance of being selected. If it isn't it is **biased**. For example, 'adults only' would be biased. (See Section 1.7 for more information on bias.)

◉ To make a sample representative, each individual in the sample should be picked at random. This process is known as taking a simple **random sample.**

◉ To take a simple random sample:

   1. Each person or item in the population is given a number.

   2. If a sample of 10 is needed, then 10 numbers are selected. This can be either: from a random number table; by a random number generator on a calculator; by using a computer; or by putting the numbers in a hat.
   The people whose numbers are selected then form the sample.

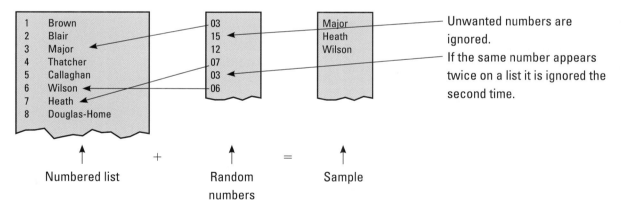

| Numbered list | + | Random numbers | = | Sample |

Unwanted numbers are ignored.
If the same number appears twice on a list it is ignored the second time.

> **Example 2**
>
> John is collecting data from each of the 50 students in his year group about the number of brothers and sisters they have.
>
> **a** Give a reason why he might use a simple random sample.
> This is an extract from a set of random numbers.
>   3352211705324821463211430592882334171412
> **b** Starting at 33 and working across, use the random numbers to give eight numbers less than 50.
> **c** Explain how John would use these numbers to take a sample of eight students.

a The number of students is large and it would take a long time to collect that data. ← This is one reason for taking a sample. Cost could be another reason.

b 33 21 17 05 32 48 46 11 ← Start at 33 and take the digits in pairs. 05 counts as the number 5. Numbers like 21 which repeat are ignored when they appear a second time. If the population was 150 you would take the digits in threes.

c He would number the students and select the ones corresponding to these numbers.

Exercise 1B

C

1 Write down two ways in which you can generate random numbers.

2 Explain what is meant by a simple random sample.

B

3 A call centre has 60 workers. Eight are to be chosen for a new training scheme.
The manager decides to choose a simple random sample of eight.
He uses a calculator to generate random numbers. These are the first few numbers he generates.
21 32 67 54 89 78 90 34 26 45 78 54 35 64 22 …
Describe how he could use these numbers to get his sample of workers.

1.3 Stratified sampling

◉ Objective

● You can select and use a stratified sample.

❓ Why do this?

A school has an equal number of boy and girl students. A simple random sample, could contain more boys than girls. A stratified sample would contain an equal number of each.

◈ Get Ready

1. Fifteen of a class of 25 students are girls.
   a What fraction are girls?   b What fraction are boys?
2. There are three classes in Year 11. There are 22 students in class A, 28 in class B and 30 in class C.
   a How many students are in Year 11?   b What fraction of the students in Year 11 are in class C?

🌐 Key Points

● A population may contain groups in which the observation of interest is likely to differ. For example, if you are looking at the heights of students then the boys' heights are likely to be different to the girls' heights. These groups are called strata (singular stratum).
● A **stratified sample** is one in which the population is split into strata, and a simple random sample is taken from each stratum. The number taken from each stratum should be in **proportion** to the total number in each stratum.

● To find the number to be selected from a stratum:

1. Find what fraction of the population is in the stratum.

Fraction in stratum $= \dfrac{\text{number in stratum}}{\text{number in population}}$

2. Multiply the fraction in the stratum by the total size of the sample.

The number sampled in a stratum $= \dfrac{\text{number in stratum}}{\text{number in population}} \times$ total sample size

Example 3

The table below shows the number in each year group of a school.
A sample of 60 students is to be taken.
How many students from each year group should be in the sample?

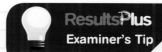
ResultsPlus
Examiner's Tip

Year	7	8	9	10	11
Number of students	150	150	100	100	100

Always make sure your individual samples total the required sample size.
Check: $15 + 15 + 10 + 10 + 10 = 60$

The sample for Year 7 will be $\frac{150}{600} \times 60 = 15$ ← In each of Years 7 and 8 there are 150 students out of 600 students.
The sample from Year 8 will be the same size as for Year 7.

The sample for Year 9 will be $\frac{100}{600} \times 60 = 10$ ← In each of Years 9, 10 and 11 there are 100 students out of 600 students.
Years 10 and 11 will also have a sample size of 10.

Exercise 1C

1 A head teacher wants to find out what Year 7 students think about their first term at their new school. He decides to ask a stratified sample of 50 students. The table shows the total number of boys and the total number of girls in Year 7.

Boys	Girls
276	324

Work out the number of boys and the number of girls he should include in the sample.

*2 A call centre allocates work according to the experience of its employees. Those with less than six months' experience do the easier work; those with more than six months' experience do more difficult tasks. There are 150 employees with less than six months' experience and 400 with more than six months' experience. Describe exactly how you would find a stratified sample of 10% of the employees.

*3 A factory owner wants to find out what his employees think about the parking facilities at his factory. He decides to ask a stratified sample of 90 of his workers. The table shows how many people are in each of the six strata he intends to use.

	Office workers	Factory floor workers	Managers
Females	50	250	10
Males	80	490	20

a Calculate the number of workers he needs to ask in each strata and describe how he should pick the individual members of each strata.

1.4 Collecting data by observation and experiment

◎ Objectives

- ○ You can interpret scales on a range of measuring instruments, including clocks.
- ○ You can design and use data collection sheets.
- ○ You can use tallying methods.
- ○ You can group data into class intervals.
- ○ You can collect data by observation, experiment or data logging.

❓ Why do this?

If you are collecting data about the number of different types of vehicles passing by, it is easier to keep a record in the form of a data collection sheet rather than trying to remember each total.

◈ Get Ready

1. What are the numbers given by each set of marks? **a** |||| **b** ||||||
2. Is there a better way of grouping the marks in part **b**?

Reading scales

◉ Key Points

- ◎ To collect data you may have to use measuring instruments.
  You must be able to read different scales on measuring instruments.
- ◎ All scales have divisions marked on them and in most cases subdivisions too.
  The scale below is a ruler.

The ruler has cm divisions, each of which is divided into ten 1 mm sections.

If an article being measured is 6 cm and 3 mm long, since 6 cm = 60 mm, the article is 63 mm long.

- ◎ Some scales are the same as the scale on the ruler. Here are a few examples.

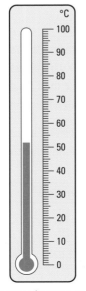

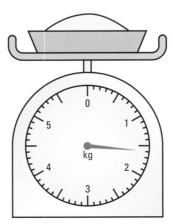

This thermometer shows a temperature of 52°C.   These scales show a weight of 1.6 kg.

Time

Key Points

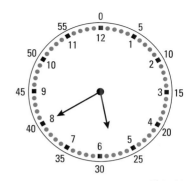

- Clocks have two scales. One shows hours and the other shows minutes.

- A clock shows 12 hours, and 60 minutes, starting from the vertical.

- There are 24 hours in the day. The 12 hour clock does not tell you if it is morning or afternoon. To tell the difference we use am for times before noon (midday) and pm for times after noon.

- Many digital clocks are 24 hour clocks.

- The 24 hour clock times always have four figures.

- Here are some examples of 12 and the corresponding 24 hour clock times.

Time	12 hour clock	24 hour clock
3 hours past midnight	3 am	03:00
Half past 11 in the morning	11.30 am	11:30
20 past three in the afternoon	3.20 pm	15:20
Quarter past 9 in the evening	9.15 pm	21:15

Example 4   What time is showing on this clock?

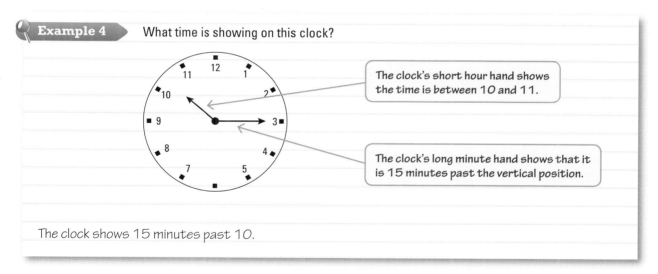

The clock's short hour hand shows the time is between 10 and 11.

The clock's long minute hand shows that it is 15 minutes past the vertical position.

The clock shows 15 minutes past 10.

Collecting data

Key Points

- There are a number of ways of collecting data.
  - **Collecting data by observation:** if you want to investigate whether a lot of traffic is caused by people taking students to school, you could observe how much traffic there is at school opening time and compare it with how much traffic there is at other times.
  - **Collecting data by experiment:** if you wish to find out how high a tennis ball bounces when dropped from different heights, you could drop a tennis ball from various heights and record how high it bounces. This would enable you to collect data on the bounce of tennis balls.

- **Data logging:** when you go to a supermarket till each item is bar coded so that it can be identified at the checkout. The number of each item sold is automatically recorded and this enables the supermarket to know what items are popular and what they need to stock up on.

When collecting data by observation a **data collection sheet** is used. The following diagram shows a data collection sheet for recording the different types of transport that might pass your door.

Vehicle	Tally	Frequency
Bicycle	ЖҀ |||	8
Bus	|||	3
Car	ЖҀ ||||	9
Lorry	ЖҀ	5
Motorcycle	ЖҀ		6
Van	ЖҀ	5

Each time a bicycle passes a **tally** mark is put next to 'Bicycle'.

When the **survey** is complete the tally marks are added together to give the total number of each vehicle. This is known as the **frequency**.

The marks are grouped into fives with the fifth tally mark drawn through the other four.

Putting tally marks in fives makes totalling up easier.

If data is numerical, and widely **spread**, you can group the data into **class intervals**. These class intervals do not have to be the same size.

When dealing with continuous data you need to make sure the intervals do not overlap. For example, the class intervals for a **variable** such as weight (w) will be of the form:

$500 \text{ g} \leqslant w < 550 \text{ g}$ This means that w is greater than or equal to 500 g but less than 550 g.

or $500 \text{ g} < w \leqslant 550 \text{ g}$ This means that w is greater than 500 g but less than or equal to 550 g.

In this case, 500 g and 550 g are the lower and upper class **limits**, while the **class size** is $550 - 500 = 50$ g.

Example 5

30 students are asked how many books they read in the Easter holiday. Each student's response is shown below. Draw and fill in a data collection sheet for this information.

1	5	9	9	6	13
6	8	4	5	5	7
9	6	11	3	8	9
14	7	7	2	12	9
3	2	1	4	0	5

ResultsPlus
Watch Out!

Make sure classes don't overlap.

Mark	Tally	Frequency
0–4	ЖҀ ||||	9
5–9	ЖҀ ЖҀ ЖҀ ||	17
10–14	||||	4
15–19		0

The marks have been grouped together into four equal class groups. The first class includes all the numbers between 0 and 4 inclusive.

In this example, different class intervals could have been chosen. For example:

Mark	Tally	Frequency
0–4	ЖҀ ||||	9
5–7	ЖҀ ЖҀ	10
8–10	ЖҀ ||	7
11–15	||||	4

Example 6　The tally chart below shows the age at marriage of a sample of men.

Age, a	Tally	Frequency
$16 < a \leqslant 20$	||	
$20 < a \leqslant 30$	ⅢⅠ ⅢⅠ |||	
$30 < a \leqslant 40$	ⅢⅠ ⅢⅠ ⅢⅠ ⅢⅠ ⅢⅠ	
$40 < a \leqslant 50$	ⅢⅠ ||	
$50 < a \leqslant 60$	|||	
$60 < a$	|	

a Fill in the frequency column.
b Write down the most popular age range in which men get married.
c Work out how many men in total there were in the sample.

a

Age, a	Tally	Frequency
$16 < a \leqslant 20$	||	2
$20 < a \leqslant 30$	ⅢⅠ ⅢⅠ |||	13
$30 < a \leqslant 40$	ⅢⅠ ⅢⅠ ⅢⅠ ⅢⅠ ⅢⅠ	25
$40 < a \leqslant 50$	ⅢⅠ ||	7
$50 < a \leqslant 60$	|||	3
$60 < a$	|	1

Add together the tallies: $5 + 5 + 3 = 13$.

Look for the class with the highest frequency.

b 30 to 40
c 51 ← Add together all of the frequencies: $2 + 13 + 25 + 7 + 3 + 1 = 51$.

Exercise 1D

1 A road traffic controller keeps a record of the types of traffic using a busy junction during a two-minute rush-hour period. This data is listed below:

Car	Car	Bus	Car	Car	Car	HGV	Bike	Car	Car	Car	Bus
Bus	HGV	Car	Car	Car	Motorbike	Bike	Car	Car	Bus	Car	HGV
Bus	Car	Car	Car	Car	Bike	HGV	Car	Car	Car	Car	Bike

a Draw a tally chart to show this data.
b Write down the name of the least common type of traffic.
c Write down the name of the most common type of traffic.

2 Here is some of the data Sally collected on the lengths, in kilometres, of different journeys.
5.6　0.86　10.5　8.654　18.49
Round all these figures to the nearest whole kilometre.

3 Here is some of the data Nassim collected on the lengths, in centimetres, of different screws.
12.63　0.96　2.54　1.52　8.45
The factory making the screws needs to know the lengths to one decimal place.
Round all these figures to one decimal place.

4 Here is part of a railway timetable.

a Which is the fastest train?

Matthew wants to travel from Preston to Birmingham. He wants to arrive before 2 pm.

b Which train should he catch?

Amelia says 'Good, this timetable shows that there is a train that arrives in Wolverhampton at 4.30 pm.'

c Explain why Amelia is wrong.

Station	Train A	Train B	Train C
Carlisle	11:09	12:07	13:00
Preston	12:17	13:17	14:28
Crewe	12:59	13:58	14:59
Wolverhampton	13:31	14:30	15:29
Birmingham	13:56	14:55	15:58

5 A shopkeeper asks 30 people entering her shop how many DVDs they have bought in the last three months. The responses are shown below:

3	5	8	9	2	7	4	10	12	3
6	2	4	9	12	13	1	7	7	11
14	3	6	5	8	1	2	7	4	3

Draw and fill in a data collection sheet showing this information. Use equal class intervals starting with the class 0−3.

6 A gardener weighs 24 tomatoes produced from plants in his greenhouse.

The weights, in grams, are shown below:

60.5	65	64.5	59	67	61.5	67	69
58	59.3	57.2	67	68.5	63	64.2	69
57	57.8	62.4	65.5	67	58	70	75

Weight (w)	Tally	Frequency
$57 \leq w < 60$		
$60 \leq w < 63$		
$63 \leq w < 66$		
$66 \leq w < 69$		
$69 \leq w$		

a Copy and complete the data collection sheet for this data.

b Write down the most common class.

c Write down the least common class.

1.5 Questionnaires

⊙ Objectives

- You can collect data by using a questionnaire.
- You can criticise questions for a questionnaire.

⊘ Why do this?

A restaurant may use a questionnaire to get feedback about its service, food and atmosphere if it is looking to make improvements.

⊕ Get Ready

Describe a good method for recording data on a data collection sheet.

Key Points

- A **questionnaire** is a list of questions designed to collect data. On questionnaires:
  - keep questions short
  - use words that are easily understood
  - do not use **biased questions** that lead the respondent to a particular answer. For example, use 'Do you agree or disagree?', rather than 'You do agree, don't you?'
  - write questions that address a single issue. For example, use 'Do you have a car?' rather than 'Do you have a petrol engine car?'
- There are two types of question to use on questionnaires.
  - An **open question** is one that has no suggested answers.
  - A **closed question** is one that has a set of answers to choose from. It is easier to summarise the data from this type of question. Closed questions will often have an opinion scale to choose from. For example:

Statistics is an important subject.				
☐	☐	☐	☐	☐
Strongly agree	Agree	Disagree	Strongly disagree	Don't know

These are **response boxes**.

This allows for other answers.

Sometimes a numerical scale is used. For example:

Tick one box to indicate your age group.				
☐	☐	☐	☐	☐
Under 20	21 to 30	31 to 40	41 to 50	Over 50

The categories do not overlap.

- When designing questionnaires, it is important to ensure that possible answers are clear, do not overlap and cover all possibilities.

Example 7

Write down what is wrong with each of these questions.

a Tick one box to indicate your age group.

☐	☐	☐	☐
Under 20	20 to 30	31 to 40	40 to 50

b How often have you had a medical in the last 4 years? Tick one box.

☐	☐	☐	☐	☐
Never	Seldom	Sometimes	Often	Very often

c Do you agree that people who have regular medicals are less likely to have major illness that goes undetected?

☐	☐
Yes	No

a The categories overlap — 40-year-olds could go into two boxes.
   Other answers are not allowed for. Where does a 60-year-old tick?

b It is difficult to decide what these words mean.

c By asking 'Do you agree ...' you are inviting the answer 'Yes'. This is called a biased question.

Exercise 1E

D A03

1 A questionnaire includes the following question.

'Do you agree that we should build a new road?'

☐   ☐

Yes   No

Write down what is wrong with this question.

2 A local council wants to know whether or not the residents would like a new swimming pool in the town. It is decided to use a questionnaire. The following questions are suggested.

A: What do you think about the idea of a new pool being built?

B: Do you want a new pool?   Yes/No

C: Where should we build a new pool?

D: Is a pool a good idea? Yes/No

Which of the above are open questions and which are closed?

C A03

3 The management of a theme park have made some changes to the amusements. They want to use a questionnaire to find out what people think about the changes. The following questions are suggested. Write down what is wrong with each of them and design a new question for each that is more suitable.

a What do you think of the new amusements?

Very good ☐   Good ☐   Satisfactory ☐

b How much money would you normally expect to pay for each amusement?

£5−£7 ☐   £7−£8 ☐   More than £8 ☐

c How often do you visit the park each year?

Often ☐   Not very often ☐

A03

*4 A supermarket manager wants to find out if people like the new layout. She decides to use a questionnaire. Write down a suitable question she could use.

1.6 Two-way tables

◎ Objectives

- You can design and use two-way tables.
- You can use information to complete a two-way table.
- You can round numbers to an appropriate degree of accuracy.

❓ Why do this?

You can use a two-way table to record results such as the drink preferences of boys and girls.

◈ Get Ready

In a class of 30 students there are:

2 left-handed girls   13 right-handed girls

4 left-handed boys   11 right-handed boys.

a How many girls are there in the class?   b How many students are left-handed?

Key Points

● Sometimes we collect two pieces of information, for example gender and eye-colour. To record this we would use a **two-way table**. A two-way table shows the frequency with which data falls into two different categories.

	Blue	Brown	Green	Total
Boys	6	14	5	25
Girls	4	16	5	25
Total	10	30	10	50

— This is the number of boys with brown eyes.
— This is the number of girls with brown eyes.
— This the total number with brown eyes.
— This is the total number of boys and girls.

● Sometimes a table is incomplete and has to be filled in before you can answer a question.

Example 8 Students in Year 11 were asked to choose their favourite drink from a choice of three. Below are the boys' and girls' responses.

A02

Girls
Tea	Coffee	Coffee	Tea	Soft
Tea	Coffee	Tea	Coffee	Tea
Soft	Soft	Tea	Tea	Soft
Coffee	Coffee	Soft	Soft	Coffee

Boys
Coffee	Coffee	Tea	Soft	Tea
Tea	Tea	Soft	Coffee	Coffee
Soft	Tea	Tea	Coffee	Coffee
Soft	Tea	Coffee	Coffee	Coffee

a Show this information in a suitable table.
b Write down the girls' top choice of drink.
c Write down the boys' top choice of drink.
d Write down the drink that was chosen by most of the students.

a

	Tea	Coffee	Soft drink	Total
Boys	7	9	4	20
Girls	7	7	6	20
Total	14	16	10	40

The most suitable table is a two-way table. Count up the number of boys that chose tea and enter it here. Do the same for the other drinks and the girls' drinks.

Total the rows and columns.

b Tea and coffee tied. ← Look for the highest number in the girls' row.

c Coffee ← Look for the highest number in the boys' row.

d Coffee ← Look for the drink which has the highest total.

Example 9 The following two-way table gives information about people's hair and eye colour.

		Eye colour			
		Brown	Green	Blue	Total
Hair colour	Brown/Black	4	4		16
	Fair	3		4	
	Ginger		1	1	4
	Total	9	8		30

ResultsPlus

Examiner's Tip

Look for rows with only one number missing and fill these in first.
The numbers in each row must add up to the row total and the same goes for columns.

a Complete the table.

b Which eye colour was most frequent?

c Which eye colour was least frequent?

a

		Eye colour			
		Brown	Green	Blue	Total
Hair colour	Brown/Black	4	4	8	16
	Fair	3	3	4	10
	Ginger	2	1	1	4
	Total	9	8	13	30

The number of blue-eyed black-haired $= 16 - 4 - 4 = 8$

The number of brown-eyed ginger-haired $= 4 - 1 - 1 = 2$

The number of green-eyed fair-haired $= 8 - 4 - 1 = 3$

The total number of fair-haired $= 3 + 3 + 4 = 10$

The total number of blue-eyed $= 30 - 8 - 9 = 13$

b Blue

c Green

 Exercise 1F

C A02

1 A number of men and women were asked which type of crisps they liked best. A total of twelve people said Plain, of which seven were men. Six women liked Salt and Vinegar. Fourteen men and twelve women liked Cheese and Onion. There were 28 men in total.

  a Draw and complete a table of the data.

  b How many people liked Salt and Vinegar crisps best?

  c How many people were asked altogether?

2 In a supermarket survey 30 men and 30 women were asked whether they preferred orange juice or grapefruit juice. 22 men preferred orange juice. 12 women preferred grapefruit juice.

  a Draw up a two-way table to show this information.

  b How many people liked orange juice best?

3 A factory employs 12 supervisors, of which 2 are female; 14 office staff, of which 3 are male; and 120 shop floor workers, of which 38 are female.
   a Draw up a two-way table to show this information.
   b Write down the number of female employees.
   c Write down the total number of employees.

Rounding numbers

Key Points

- If the number after the place you want to round to is 5 or more, you round up.
- If the number after the place you want to round to is less than 5, you round down.
- You can round to a power of 10, the nearest whole number or to a given number of decimal places.
- When rounding to the nearest whole number the measurement given will be inaccurate by up to half in either direction.
  For example, a reading given as 4 could be between 3.5 and 4.5.
- To round numbers to a given number of significant figures (s.f.), you count that number of digits from the first non-zero digit. If the next digit is 5 or more then you round up. If the next digit is 4 or less you round down.
- Leading zeros in decimals are not counted as significant.

Example 10
   a Write 9736 correct to the nearest 1000.
   b Round 9736 correct to the nearest 100.
   c Round 9736 correct to the nearest 10.
   d Round 5.28 to the nearest whole number.

a The answer is 10 000. ← *The number after the thousands is 7, which is greater than 5, so you round up to 10 000.*

b The answer is 9700. ← *The number after the hundreds is 3, which is less than 5, so you round down.*

c The answer is 9740. ← *The number after the tens is 6, which is greater than 5, so you round up.*

d 5.28 rounds to 5 to the nearest whole number. ← *The number after the decimal point is a 2, which is less than 5, so you round down.*

Example 11
   a Round 3.475 to one decimal place.
   b Round 5.763 to 2 decimal places.
   c Round 2.865 to 2 decimal places.

a 3.475 rounds to 3.5 to 1 decimal place. ← *The number in the second decimal place is a 7, which is greater than 5, so round up.*

b 5.762 rounds to 5.76 to 2 decimal places. ← *The number in the third decimal place is a 2, which is less than 5, so round down.*

c 2.865 rounds to 2.87 to 2 decimal places. ← *The number in the third decimal place is a 5, so round up.*

Example 12 Round the following numbers correct to

    a 3 significant figures   **b** 2 significant figures

    i 4.7084   **ii** 0.006 375

a **i** 4.71 (3 s.f.) ←   The 8 means the 0 will be rounded up to a 1.

    ii 0.006 38 (3 s.f.) ←   The 5 means the 7 will be rounded up to an 8.

b **i** 4.7 (2 s.f.)

    ii 0.0064 (2 s.f.) ←   The 7 means the 3 will be rounded up to a 4.

Exercise 1G

1 The table gives some information about the cost of holidays in Greece.

Half Board	3 nights		1 week		10 nights		2 weeks	
Month(s) of Holiday	**Adult**	**Child**	**Adult**	**Child**	**Adult**	**Child**	**Adult**	**Child**
October 2009	340	250	469	325	569	450	729	640
Nov/Dec 2009	225	180	315	255	385	300	499	360
January 2010	215	180	315	250	375	290	529	460

Mr and Mrs Caput and their 8-year-old son Aaron decide to go on one of these holidays. They wish to go for 10 nights in November.

a Work out the total cost of the holiday to the nearest £10.

b Discuss how they could have reduced the cost of their holiday.

1.7 Sources of bias

◎ Objectives

- You can identify possible sources of bias.
- You understand how different sample sizes may affect the reliability of any conclusions drawn.

? Why do this?

If you want to accurately estimate the average height of students you need to collect reliable data. For example, if you include more boys than girls in your sample then you are likely to get a taller average.

Key Points

- When collecting data you should make sure that the data is representative of the population it is taken from. Data that does not do this is said to be biased. There are several types of bias.

Selection bias

- If you select only people who shop at a supermarket and ask them what they think about how that supermarket compares with a rival supermarket, you will get a biased opinion, since the people who use the other supermarket are not represented. This is called under-coverage bias.

- If you ask people to fill in a questionnaire and post it back to you, only a certain type of person will bother to respond. The respondents will not be representative of the general public. This is non-response bias.
- If you ask people to text a radio show about a controversial topic, you will get mainly people who have a strong opinion about that topic. This is voluntary response bias.
- Selection bias can be avoided by random selection and random allocation.

Measurement bias

- If you ask people if they are satisfied, dissatisfied or very dissatisfied, you are likely to get a biased opinion because there are two answers for dissatisfied and only one for satisfied.
- If you ask a question such as, 'You are satisfied, aren't you?' you are more likely to get 'Yes' as an answer. These are called **leading questions**.
- People like to present themselves in the best light, so if you ask them, 'Do you often behave unreasonably?' you are likely to get 'No' for an answer even if it should be 'Yes'.

Sampling error

- If you take two random samples you are unlikely to get exactly the same result (though they should be close to each other). This is called the sampling error.
- Increasing the sample size will reduce the sampling error. It is difficult to say how big a sample should be as this depends on how varied the population is. However, the larger the sample the more representative it will be of the population and the more accurate the information will be.

Example 13 Write down, with reasons, whether or not each of the following is biased.

A03

a You want to find out what people think about a football team. You ask supporters as they enter the ground before a match.

b You wish to find out what proportion of the population has had flu in the last month. You interview people in the doctor's waiting room.

c You ask the first 10 people you meet, 'Do you agree that banning smoking in public places is a good thing?'

d You ask, 'Have you ever been convicted of drink driving?'

e You ask three people what they thought of the Eurovision Song Contest.

a Biased. Non-supporters are not represented.

b Biased. People who do not visit the doctor are not represented.

c Biased. Not everyone has an equal chance of being asked.
The question is leading the respondent to agree.

d Biased. This is a sensitive question. You are not likely to get a true answer.

e Biased. The sample is too small.

Exercise 1H

1 An examination board wants to get information on schools' views regarding how they respond to queries. They send a questionnaire to a sample of schools in the London area. Is this a biased sample? Give one reason for your answer.

C

C

2 Write down, with reasons, whether or not each of the following are biased.

A: A hospital wants to know how often people use A & E. They ask all the people attending A & E on one particular Wednesday.

B: An opinion poll company wants to find out how voters would vote if there were to be an election next week. They conduct a telephone poll of 20 voters in each of 10 towns.

C: A manufacturer of climbing ropes wants to see if his ropes are of the strength he advertises. He tests a sample. He tests every tenth rope made.

D: You ask 50 people using a recycling facility what they think about recycling.

A03

*3 One hundred people attend a rally on 'action for climate change'.

David says, 'That is a lot of people. They must be right.'

Jody says, 'I disagree.'

Discuss the views of David and Jody.

1.8 Secondary data

◎ Objective

● You can extract data from lists and tables.

◈ Why do this?

You want to find out how many accidents there were in your town last year. You can't count these yourself so you have to get the data from published sources.

◈ Get Ready

If you needed a new mobile phone, where would you look to find one that best suited you at a price you could afford?

Key Points

● Secondary data can be obtained relatively quickly and cheaply from a number of sources, including reference books, journals, newspapers and the internet. Remember, however, that the data may be inaccurate or out of date. Only use data from a reliable source and check the data against another source if possible.

● It is also possible to obtain secondary data from a database. A **database** is an organised collection of information, usually stored on a computer.

● The spreadsheet below shows part of a database kept on a computer. The entries at the top in red are fields. The entries below in black are the records. They can be easily changed to be arranged in numerical, alphabetical, gender or age order.

ID number	Surname	Forename	Gender	Age
01	Abbot	David	M	32
02	Adair	Jakie	F	27
03	Allison	Paul	M	45
04	Barber	Hassan	M	25
05	Baxter	Jenny	F	38

Example 14 Part of a database for second-hand Ford Mondeo cars is shown below.

Vehicle summary	Colour	Engine	Mileage	Price	Year
Ford Mondeo Edge	Black	2000cc petrol	11 549	£9 995	2006
Ford Mondeo Edge	Blue	2000cc petrol	14 100	£10 499	2008
Ford Mondeo Edge	Grey	2000cc petrol	10 400	£11 599	2008
Ford Mondeo Edge	Grey	2000cc petrol	12 654	£11 494	2007
Ford Mondeo Edge	Blue	2000cc petrol	7520	£11 999	2008
Ford Mondeo Zetec	Silver	2000cc petrol	10 078	£11 995	2008
Ford Mondeo Zetec	Silver	2000cc petrol	12 088	£14 995	2008
Ford Mondeo Titanium	Grey	2000cc petrol	11 555	£12 395	2008
Ford Mondeo Zetec	Black	2000cc petrol	5800	£12 895	2008
Ford Mondeo Zetec	Silver	2000cc petrol	12 123	£12 995	2008

a Which four fields could be used to order the data?

b What was the mileage of the car that cost over £13 000?

c What colour was the car that had driven the least number of miles?

d What was the maximum mileage driven by one of these cars?

a Mileage, price, year, colour. ← Mileage price and year could be put in numerical order. Colour could be put in alphabetical order.

b 12 088 ← Find the car that had a price greater than £13 000 and look in its mileage column.

c Black ← In the mileage column, find the car that had driven the least mileage then look across its row to the colour column.

d 14 100 ← Look for the largest number in the mileage column.

Exercise 1I

1 The following database gives some information about the CO_2 emissions, in thousand tonnes of carbon dioxide equivalent, in a certain country.

	Year					
	2003	2004	2005	2006	2007	2008
Buses	323	344	355	342	394	421
Cars	6280	6251	6163	6159	6063	6055
HGVs	2147	2154	2235	2295	2162	2221
Motorcycles	39	41	44	40	41	39
Railways	231	209	225	241	245	251

a What were the emissions for motorcycles in 2007?

b Which form of transport produced the most emissions?

c Write down the year when emissions for railways were lowest.

d Write down the method of transport for which the emissions have dropped each year.

2   The following database gives information about the weather in a certain town during the first six months of the year.

	Max temp °C	Min temp °C	Air frost days	Sunshine hours	Rainfall mm	Days of rainfall ⩾ 1 mm
January	6.4	1.2	10.7	44.3	101.9	15.3
February	6.9	1.3	9.6	72.0	73.4	11.3
March	8.8	2.5	6.3	107.9	78.3	14.1
April	11.4	3.5	3.8	155.1	50.7	10.6
May	15.0	6.1	1.0	214.8	55.0	10.0
June	17.1	9.0	0	197.7	67.9	10.7

a   How many days of air frost were there in March?

b   Write down the month that had the least number of days of rainfall.

c   Which was the sunniest month?

d   Which month had the greatest difference between maximum and minimum temperatures?

3   The database below is part of an agricultural survey of cattle in England in 2006, 2007 and 2008. The numbers of cattle are given in thousands.

	Year		
	2006	2007	2008
Female cattle			
Aged 2 years or more	2550	2531	2475
Total breeding herd (cattle that have calved)	2043	2027	1994
Beef	767	768	758
Dairy	1276	1259	1236
Other female cattle (not calved)	507	504	481
Beef	224	220	216
Dairy	293	284	265
Aged between 1 and 2 years	825	799	778
Beef	512	502	497
Dairy	313	297	282
Male cattle			
Aged 2 years or more	217	217	217
Aged between 1 and 2 years	619	583	578

a   Write down the number of female cattle aged between one and two years in 2007.

b   What do you notice about the numbers of male cattle aged two years or more throughout the three years?

c   Were there more female beef or more female dairy cattle in 2008?

d   What conclusions can you draw about the trend in the numbers of cattle over the three years?

Chapter review

- **Statistics** is used to provide information. The statistical problem-solving process can be shown as a simple diagram:

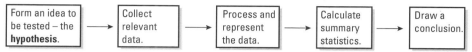

- **Primary data** is data you collect yourself.
- **Secondary data** is data that has been collected by others.
- **Qualitative data** can be described in words.
- **Quantitative data** are numerical observations.
- **Discrete data** can only take certain numerical values.
- **Continuous data** can take any numerical value.
- A **sample** is part of a population that is used to give information about the population as a whole in a **sample survey**. The sample must be representative of all the people or items being investigated, with each member of the population having an equal chance of being selected.
- A simple **random sample** is one where each person is given the same chance of being included.
- A **stratified sample** is one in which the population is split into groups called strata and a simple random sample is taken from each stratum. The number taken from each stratum is proportional to the size of the stratum.
- Data can be collected by **observation, experiment** or **data logging**.
- When collecting data by observation a **data collection sheet** is used.
- When dealing with continuous data you need to make sure the intervals do not overlap.
- If data is numerical, and widely spread, you can group the data into **class intervals**. These class intervals do not have to be the same size.
- A **questionnaire** is a list of questions designed to collect data.
- An **open question** is one that has no suggested answers.
- A **closed question** is one that has a set of answers to choose from.
- When designing questionnaires, it is important to ensure that possible answers are clear, do not overlap and cover all possibilities.
- A **two-way table** shows the frequency with which data falls into two different categories.
- **Biased** data is data that does not represent the population that it is taken from.
- A **database** is an organised collection of information.

Review exercise

1. a In an experiment to look at the growth rate of beans Phoebe measures the height of 5 bean shoots.
      The heights are 6.7 cm, 7.5 cm, 5.5 cm, 6.3 cm and 7.2 cm.
      Write these measurements to the nearest whole number.

   b James uses a measuring instrument to measure the lengths of 5 bolts. The lengths are 4.56 cm, 3.98 cm, 4.55 cm, 5.67 cm and 3.95 cm.
      Write these measurements to the nearest one decimal place.

D

2 James wants to find out how many text messages people send.
He uses this question on a questionnaire.

'How many text messages do you send?'

1 to 10 ☐   11 to 20 ☐   21 to 30 ☐   more than 30 ☐

a Write down **two** things wrong with this question.

James asks 10 students in his class to complete his questionnaire.

b Give **one** reason why this may not be a suitable sample.   *March 2009*

3 Poppy wants to find out how much time people use their computer for.
She uses this questionnaire.

For how much time do you use your computer?

0–1 hours ☐   3–4 hours ☐

1–2 hours ☐   4–5 hours ☐

2–3 hours ☐   5–6 hours ☐

a Write down **two** things that are wrong with this question.

Poppy gives her questionnaire to all the students in her class. Her sample is biased.

b Give **one** reason why.   *Nov 2008*

A02
A03

4 Courtney lives in Oxenholme.
She has an interview for a job in Glasgow at 10.30 am. The interview will take 45 minutes.
She will travel by train.
It takes Courtney 20 minutes to walk from her home to the station.
It takes 35 minutes for Courtney to walk from the station to the place of her interview.
Here is part of the train timetable from Lancaster to Glasgow and from Glasgow to Lancaster.

Lancaster to Glasgow			
Lancaster	06:54	08:08	08:29
Oxenholme	07:08	08:22	08:43
Penrith	07:34		09:19
Carlisle	07:50	09:01	09:21
Glasgow	09:14	10:16	10:46

Glasgow to Lancaster			
Glasgow	14:40	16:00	16:40
Carlisle	15:49	17:09	17:51
Penrith			18:05
Oxenholme	16:24	17:42	18:29
Lancaster	16:38	17:54	18:44

She goes straight to the interview.
She wants to do some shopping in Glasgow but wants to catch the quickest train home after 2 pm.

a Plan a schedule for Courtney.

The fastest train from Lancaster to Glasgow in the morning is the 08:08.

b Suggest reasons for this.

	Time
Courtney leaves home	
Train departs Oxenholme	
Train arrives Glasgow	
Courtney arrives for interview	
Interview finished	
Train leaves Glasgow	
Train arrives Oxenholme	
Courtney arrives home	

5 The table shows some information about the cost, in £s, of all inclusive holidays to Bahrain. The price is per adult. There is a 20% reduction for children.

Hotel	Economy Class		Business Class	
	3 nights	5 nights	3 nights	5 nights
Metro	469	595	1219	1345
Habtoor	505	655	1350	1505
Hilton	510	659	1259	1410
Atlantis	659	975	1469	1735

Wing, his wife and 12-year-old daughter plan to go to the Hilton Hotel for 5 nights, travelling Business Class.

a How much will Wing have to pay for his family to take this trip?

b Discuss ways in which Wing could reduce the cost of this holiday.

6 Naomi wants to find out how often adults go to the cinema.
She uses this question on a questionnaire.

> 'How many times do you go to the cinema?'
>
> ☐   ☐   ☐
>
> Not very often   Sometimes   A lot

a Write down **two** things wrong with this question.

b Design a better question for her questionnaire to find out how often adults go to the cinema. You should include some response boxes.

Nov 2008

*7 Valerie is the manager of a supermarket.
She wants to find out how often people shop at her supermarket.
She will use a questionnaire.

Design a suitable question for Valerie to use on her questionnaire.
You must include some response boxes.

June 2008

*8 Yolande wants to collect information about the number of e-mails the students in her class send.
Design a suitable question she could use on a questionnaire.
You must include some response boxes.

March 2008

9 Melanie wants to find out how often people go to the cinema.
She gives a questionnaire to all the women leaving a cinema.

Her sample is biased.
Give **two** possible reasons why.

March 2008

*10 Amberish is going to carry out a survey about zoo animals.
He decides to ask some people whether they prefer lions, tigers, elephants, monkeys or giraffes.

Design a data collection sheet that he can use to carry out his survey.

March 2006

C A02 A03

* **11** Angela asked 20 people in which country they spent their last holiday.
Here are their answers.

France	Spain	Italy	England
Spain	England	France	Spain
Italy	France	England	Spain
Spain	Italy	Spain	France
England	Spain	France	Italy

Design **and** complete a suitable data collection sheet that Angela could have used to show this
information.

March 2004

A03

12 The manager of a Country Park asks the following two questions on a questionnaire.

'Do you go to the Country Park?' Sometimes ☐ Often ☐

'How old are you?' 0 to 10 years ☐ 10 to 20 years ☐ Over 20 years ☐

a What is wrong with each of these questions?

b For both questions above, write a better version that the manager can use.

A03

13 Write down, with reasons, whether or not each of the following is biased.

a A call centre manager wants to know how easy it is to use the staff reference sheets when
answering a call. He asks all the people working on the night shift.

b A mobile phone company wants to find out what people think about their new pricing contract and
randomly select 10% to ask.

c A town council poses the question 'Do you agree that we are doing a good job in the area of
recycling?'

A

14 The two-way table shows information about the number of students in a school.

	Year Group					Total
	7	8	9	10	11	
Boys	126	142	140	135	127	670
Girls	134	140	167	125	149	715
Total	260	282	307	260	276	1385

Robert carries out a survey of these students.
He uses a sample of 50 students stratified by gender and by year group.
Calculate the number of girls from Year 9 that are in his sample.

June 2008

15 The table shows the number of boys in each of four groups.

Group	A	B	C	D	Total
Number of boys	32	41	38	19	132

Jamie takes a sample of 40 boys stratified by group.
Calculate the number of boys from group B that should be in his sample.

March 2008

16 258 students each study one of three languages.
The table shows information about these students.

	Language studied		
	German	French	Spanish
Male	45	52	26
Female	25	48	62

A sample, stratified by the language studied and by gender, of 50 of the 258 students is taken.

a Work out the number of male students studying Spanish in the sample.

b Work out the number of female students in the sample.

June 2009

17

	Male	Female
First year	399	602
Second year	252	198

The table gives information about the numbers of students in the two years of a college course. Anna wants to interview some of these students. She takes a random sample of 70 students stratified by year and by gender.

Work out the number of students in the sample who are male and in the first year.

Nov 2008

ResultsPlus
Exam Question Report

82% of students answered this sort of question poorly.

* **18** 80 children went on a school trip.
They went to London or to York.
23 boys and 19 girls went to London.
14 boys went to York.
Draw and complete a suitable table of this information.

March 2009

A

A02 A*

2 AVERAGES AND RANGE

BBC Video

On average, the Rhode Island Red chicken lays 275 eggs per year. Not all chickens of this breed, however, will lay exactly this number of eggs. What do we mean by average and how do we allow for different chickens laying different numbers of eggs?
When you have read this chapter you will be able to answer such questions.

◉ Objectives

In this chapter you will:
- use number operations with whole, negative and decimal numbers
- learn about mean, mode and median
- use algebraic notation
- consider the advantages and disadvantages of using each of these three measures of average
- use a calculator to solve complex calculations
- look at range, quartiles and interquartile range.

◈ Before you start

You should know that:
- quantitative data are data that can be written as a number
- frequency is the number of times something occurs.

2.1 Understanding and using numbers

Objectives

- You can add, subtract, multiply and divide any number.
- You understand and use number operations and the relationship between them, including inverse operations and the hierarchy of operations.

Why do this?

This will help you to solve problems and to calculate statistics such as the mean of a set of data.

Get Ready

Work out

a $27 + 45$ **b** $116 - 51$

Adding and subtracting decimals

Key Points

- When adding or subtracting numbers, keep the decimal points in a line.
  In whole numbers you have to imagine the **decimal point** after the last digit.

Example 1

Graham measures 3 pieces of metal pipe.

They measure 1.62 m, 1.56 m and 1.23 m each.

Work out the total length of the pipes.

```
  1 . 6 2
+ 1 . 5 6
+ 1 . 2 3
  4 . 4 1
```

← Write the numbers under each other, making sure the decimal points line up.

← Add the numbers.

Total length = 4.41 m

The decimal point in the answer goes under the other decimal points.

Example 2

Melita has 15.83 litres of petrol in her car.

She goes on a trip and uses 4.91 litres of petrol.

Work out how much petrol she has in her car.

```
  1 5 . 8 3
-   4 . 9 1
  1 0 . 9 2
```

Write the numbers under each other, making sure the decimal points line up.

← Subtract the numbers.

Amount of petrol left = 10.92 litres

The decimal point in the answer goes under the other decimal points.

Multiplying and dividing decimals

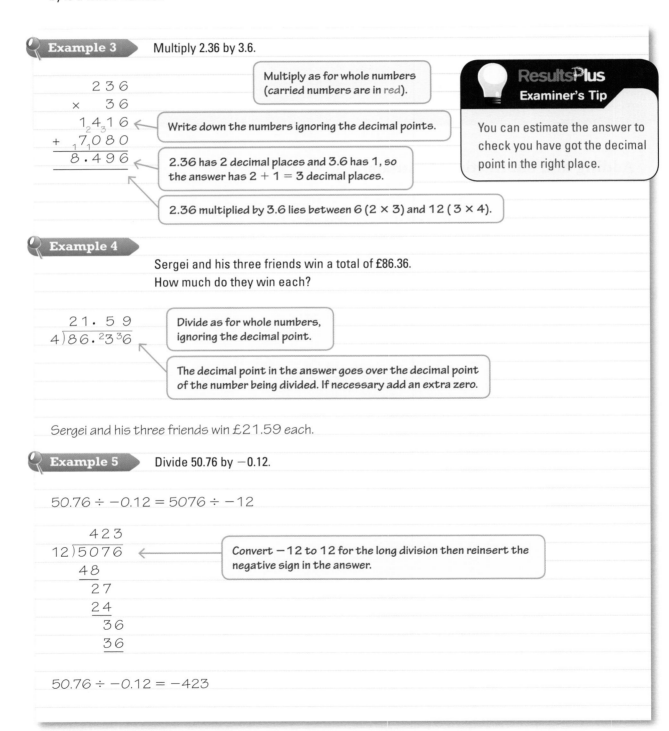

Key Points

- When multiplying decimals the number of decimal places in the answer is the sum of the number of decimal places in the numbers being multiplied.
- When dividing by a decimal number, continually multiply both numbers by 10 until the number you are dividing by is a whole number.

Example 3   Multiply 2.36 by 3.6.

```
      2 3 6
    ×   3 6
    1 4 1 6
  + 7 0 8 0
    8 . 4 9 6
```

Multiply as for whole numbers (carried numbers are in red).

Write down the numbers ignoring the decimal points.

2.36 has 2 decimal places and 3.6 has 1, so the answer has 2 + 1 = 3 decimal places.

2.36 multiplied by 3.6 lies between 6 (2 × 3) and 12 (3 × 4).

ResultsPlus
Examiner's Tip

You can estimate the answer to check you have got the decimal point in the right place.

Example 4

Sergei and his three friends win a total of £86.36.
How much do they win each?

```
    2 1 . 5 9
  4)8 6 . ²3 ³6
```

Divide as for whole numbers, ignoring the decimal point.

The decimal point in the answer goes over the decimal point of the number being divided. If necessary add an extra zero.

Sergei and his three friends win £21.59 each.

Example 5   Divide 50.76 by −0.12.

$50.76 \div -0.12 = 5076 \div -12$

```
        4 2 3
  12)5 0 7 6
      4 8
      2 7
      2 4
        3 6
        3 6
```

Convert −12 to 12 for the long division then reinsert the negative sign in the answer.

$50.76 \div -0.12 = -423$

Powers and roots of numbers

Key Points

* Expressions like $2 \times 2 \times 2$ can be written in a short form as 2^3. The three is called the **power** or index and it tells you how many times the 2 is to be multiplied by itself.
* $3 \times 3 = 9$ but $-3 \times -3 = 9$, so the **square root** ($\sqrt{}$) of 9 has two possible values, 3 and -3.
  -3 is called the negative square root of 9.
  3 is called the positive square root of 9.
* In the same way the **cube root** ($\sqrt[3]{}$) of 1000 is 10 (since $10^3 = 1000$).
* $-10 \times -10 \times -10 = -1000$ so this number has only one cube root.
* Alternatively, the cube root can be written as an index. $\sqrt[3]{1000}$ is written as $1000^{\frac{1}{3}}$.

Example 6 Work out

a 2^3 b 10^3 c $\sqrt[3]{64}$

a $2^3 = 2 \times 2 \times 2 = 8$

> Multiply the number together the number of times given by the index.

b $10^3 = 10 \times 10 \times 10 = 1000$

c $\sqrt[3]{64} = 4$

> $4 \times 4 \times 4 = 64$

Order of operations

Key Points

* You use a set of rules to tell you the order in which operations are to be done.
  The rules can be remembered using the word **BIDMAS** which gives the order in which the operations are carried out.
  Brackets
  Indices
  Division
  Multiplication
  Addition
  Subtraction

Example 7 Work out

$$4 \times 3^2 + (8 - 4)$$

$(8 - 4) = 4$, so $4 \times 3^2 + (8 - 4) = 4 \times 3^2 + 4$
$3^2 = 9$, so $4 \times 3^2 + 4 = 4 \times 9 + 4$
$4 \times 9 = 36$, so $4 \times 9 + 4 = 36 + 4 = 40$

> Brackets first
> Indices next
> There is no Division, so Multiplication comes next, followed by Addition.
> There is no Subtraction.

Example 8 Work out

    a $(4^2 + 3^3) \times 2$

    b $\dfrac{3 \times 4 + 4}{3 + 1}$

a $4^2 + 3^3 = 16 + 27 = 43$ ← Brackets first.

    $43 \times 2 = 86$ ← No Indices and no Division, so Multiply next.

    $(4^2 + 3^3) \times 2 = 86$

b $\dfrac{3 \times 4 + 4}{3 + 1} = (3 \times 4 + 4) \div (3 + 1)$ ← Get rid of fraction form.

    $(3 \times 4 + 4) = 12 + 4 = 16$ ← Brackets first.

    $(3 + 1) = 4$

    $16 \div 4 = 4$ ← No Indices and so Division next.

    $\dfrac{3 \times 4 + 4}{3 + 1} = 4$

Multiplying and dividing negative numbers

Key Points

⦿ When two positive numbers are multiplied or divided, the result will be a **positive number**.

⦿ When a positive and a negative number are multiplied or divided, the result will be a **negative number**.

⦿ When two negative numbers are multiplied or divided, the result will be a positive number.

⦿ This is easily remembered.
  ⦿ If the signs are the same it is positive.
  ⦿ If the signs are different it is negative.

Example 9 Work out

    a $(+2) \times (-6)$

    b $(-2) \times (+3)$

    c $(-8) \div (-2)$

    d $(+10) \div (-2)$

a $(+2) \times (-6) = (-12)$ ← The signs are different so negative.

b $(-2) \times (+3) = (-6)$ ← The signs are different so negative.

c $(-8) \div (-2) = (+4)$ ← The signs are the same so positive.

d $(+10) \div (-2) = (-5)$ ← The signs are different so negative.

Exercise 2A

Questions in this chapter are targeted at the grades indicated.

1   An electrical shop keeps records of every till transaction.
    The following are the amounts of money taken in the first hour on one Saturday morning.

    £125.00, £344.90, £25.56, £76.00, £62.50, £545.55

    a Work out how much the shop takes in this hour.

    b The till contains £20.40 of 10p coins. How many 10p coins is this altogether?

    c If the shop owner changes as many of the 10p coins as possible into 50p coins, how many 50p coins
       will he have?

2   Naomi wants to do an experiment. She has a piece of wood 2.88 metres long.
    If she divides the piece of wood into 24 equal size pieces, how long will each piece be?

3   Write down the answers to the following.

    a 9^2 b 5^3 c $\sqrt{169}$

    d $\sqrt[3]{343}$ e $5^2 - (6 - 3) - (4 + 1)$ f $7 + 6 \times 4 - (10 \div 2.5)$

4   Max employs a painter to paint the walls and ceilings of 3 rooms. They are each 3.5 m long, 4 m wide
    and 2 m high. The paint for 1 m^2 costs £0.85. The labour costs £110 per room.
    Max takes £450 out of his savings account to pay the painter.
    Investigate whether or not this will be enough money to employ the painter.

5   George has 8.50 metres of wood. He needs 2.65 metres to make an owl box and 1.45 metres to make a
    robin box. He wants to make at least one owl box.
    Investigate how many of each box he could make to leave the least waste.

D

A02
A03

A02
A03

2.2 Finding the mode and median

◎ Objective

◎ You can find the mode and the median of a set of data.

❓ Why do this?

You could work out the middle or average value of
the students' heights in your class to find out who
is taller or shorter than average.

⬆ Get Ready

Arrange the following sets of numbers in order, starting with the smallest number and ending with the largest.

a 2   5   3   8   12   4   10   5   7   9   1   8   b 3.5   6.2   4.5   8.7   12.5   4.6   3.5

🔍 Key Points

◎ A quantitative data set is often described by giving a single value that is representative of all the values in
the set. We call this value the **average**. For example, we might say, 'The average number of matches in a
matchbox is 50'. There are three different measures of average commonly used:

  ◎ the **mode**
  ◎ the **median**
  ◎ the **mean**.

This section introduces the first two of these measures.

Key Points

◉ The mode of a set of discrete data is the value that occurs most frequently.
  ◉ It may not exist, if all values occur with exactly the same frequency.
  ◉ When it does exist, it will always be one of the observations.
  ◉ There may be more than one mode.
  ◉ It could be the smallest or largest value.
◉ The median is the middle value when the data are ordered from the smallest to the largest.
  It splits the data into two

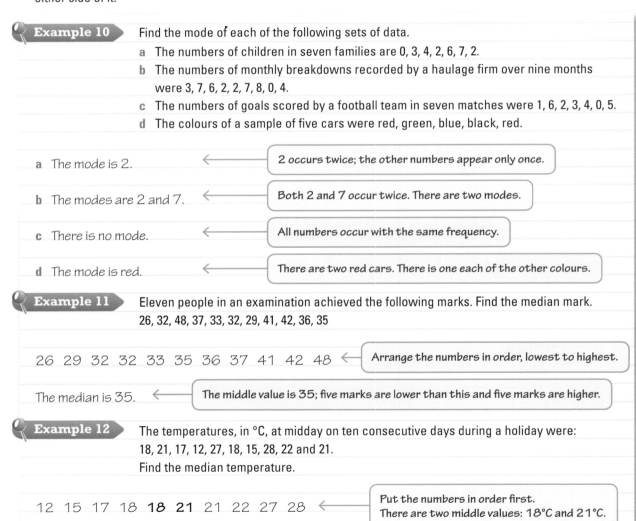

Lowest value   Median   Highest value

50% of data   50% of data

◉ If there are two middle values in a set of data the median is halfway between them.
  If there are n observations, add one to n then divide by 2. If this is a whole number the median is the value
  of this term; if it is not a whole number the median is midway between the values of the two whole numbers
  either side of it.

Example 10 ▸ Find the mode of each of the following sets of data.
  a The numbers of children in seven families are 0, 3, 4, 2, 6, 7, 2.
  b The numbers of monthly breakdowns recorded by a haulage firm over nine months
    were 3, 7, 6, 2, 2, 7, 8, 0, 4.
  c The numbers of goals scored by a football team in seven matches were 1, 6, 2, 3, 4, 0, 5.
  d The colours of a sample of five cars were red, green, blue, black, red.

a The mode is 2.   ← | 2 occurs twice; the other numbers appear only once. |

b The modes are 2 and 7. ← | Both 2 and 7 occur twice. There are two modes. |

c There is no mode.   ← | All numbers occur with the same frequency. |

d The mode is red.   ← | There are two red cars. There is one each of the other colours. |

Example 11 ▸ Eleven people in an examination achieved the following marks. Find the median mark.
  26, 32, 48, 37, 33, 32, 29, 41, 42, 36, 35

26 29 32 32 33 35 36 37 41 42 48 ← | Arrange the numbers in order, lowest to highest. |

The median is 35. ← | The middle value is 35; five marks are lower than this and five marks are higher. |

Example 12 ▸ The temperatures, in °C, at midday on ten consecutive days during a holiday were:
  18, 21, 17, 12, 27, 18, 15, 28, 22 and 21.
  Find the median temperature.

12 15 17 18 **18 21** 21 22 27 28 ← | Put the numbers in order first. There are two middle values: 18°C and 21°C. |

The median is $\dfrac{18 + 21}{2} = 19.5\,°C.$ ← | Halfway between 18 and 21 is 19.5 °C. |

Exercise 2B

1. During a football season a school team played 30 matches.
They scored the following numbers of goals in each game.

1	6	7	2	3	3	0	6	3	2
0	1	5	3	0	2	1	4	0	6
2	5	1	0	2	2	4	0	1	2

Find the mode.

2. A salesman buys fuel every day at his local garage.
The following are the quantities he buys in one particular week.

Monday 20 litres   Tuesday 14 litres   Wednesday 16 litres   Thursday 12 litres
Friday 17 litres   Saturday 12 litres   Sunday 30 litres

Find the median amount he buys.

3. Listed below are the ages of the members of a girls' table tennis club.

17	16	18	16	17	15	18	16	18
17	18	17	16	18	17	16	17	17
16	14	16	16	15	15	14

a Find the mode.   b Find the median age.

4. A train travelling from London to Glasgow makes seven stops.
The following figures show the number of passengers on each stage of the journey.

London to Milton Keynes   240   Milton Keynes to Crewe   250
Crewe to Warrington   220   Warrington to Preston   234
Preston to Lancaster   190   Lancaster to Penrith   170
Penrith to Carlisle   180   Carlisle to Glasgow   245

Find the median number of passengers on the train.

2.3 Algebra

Objectives

- You can use notation and symbols correctly.
- You can distinguish between the words 'equation', 'formula' and 'expression'.

Why do this?

By using letters to represent numbers you can write down general expressions and formulae that can be applied to a variety of situations.
You could, for example, write down a general formula for the mean of a set of numbers.

Get Ready

1. 3 apples + 2 apples = ... apples
2. $a + a + a + a + a =$

Using letters to represent numbers

Key Points

◉ A letter can be used in place of a number.

Example 13 There are 5 red balls and n white balls in a bag.

Three balls are drawn from the bag, 2 red and 1 white.
a How many red balls are there left in the bag?
b How many balls are there left altogether?

The bag now has 5 red balls added and a further n white balls.
c How many balls are there in the bag?

a There are $5 - 2 = 3$ red balls left in the bag.

b There are $5 + n - 3 = 2 + n$ balls left in the bag.

c There are now $2 + n + 5 + n = 7 + 2n$ balls in the bag.

Example 14 Rageh buys 10 cakes costing x pence each and a sandwich costing 98 pence.
Write down in terms of x how much Rageh spent.

Rageh spent $10x + 98$ pence.

Expressions, identities, equations and formulae

Key Points

◉ $10x + 98$ is called an **algebraic expression**.
An algebraic expression contains at least one letter.

◉ An **identity** has the left-hand side the same as the right-hand side.
$5 + n - 3 = 2 + n$ is an identity because the two sides are identical.

◉ An **equation** also contains at least one letter but also contains an $=$ sign.
$23x - 6 = 0$ is an equation.

◉ An equation can be solved to give a value for the unknown amount represented by the letter.
If $23x - 6 = 0$ then $23x = 6$ and $x = \frac{6}{23}$.

◉ A **formula** has at least two letters in it, and given a value for one of the letters the value of the other letter can be found.
$y = 2x + 4$ is a formula.
When $x = 1$, $y = 2 \times 1 + 4 = 2 + 4 = 6$
When $x = 2$, $y = 2 \times 2 + 4 = 4 + 4 = 8$
You could find a value of y for any given value of x.

algebraic expression   identity   equation   formula

Example 15 Write down whether each of these is an expression, an identity, an equation or a formula.

a $3x + 4$
b $x = 4y + c$
c $x + 4 = 5x$
d $2(x - 3) = 2x - 6$

a This is an expression. ← It has no equals sign.

b This is a formula. ← It has an equals sign and three letters.

c This is an equation. ← It has an equals sign and one letter.

d This is an identity. ← The two sides are identical.

Exercise 2C

1 Justin collects x pieces of data from each of his six friends. How many pieces of data does he have?

2 There are 7 red beads in a box and p blue beads.
Three red beads and one blue bead are taken out of the box.
a How many red beads are left in the box?
b How many beads altogether are left in the box?

3 Harry collects x pieces of data from Sarah and y pieces of data from Charlie.
Write down an expression for the number of pieces of data he has.

4 Yasmin earns p pence for every job she does. She does s jobs. Write down an expression for the total amount of money she earns.

5 Write down whether each of the following is an expression, an identity, a formula or an equation.
a $4x + 9$
b $y + 6 = 14$
c $y = 3x + 9$
d $5x + 2 - 6x = 2 - x$

6 Molly collects data on the number of cows (x) on the 6 farms surrounding her village.
She knows that $\bar{x} = 58$. Use $\bar{x} = \dfrac{\Sigma x}{n}$ to find the total (Σx) number of cows on the farms.

7 Joe buys 3 bunches of mixed flowers each containing x roses and then he picks 7 extra roses from his garden.
a Write down in terms of x how many roses he has.
Jenny buys 3 bunches of flowers each costing y pounds and a vase costing £6.00
Keith buys 3 bunches of flowers each costing $2y$ pounds each and a vase costing £3.00.
They both spend the same amount of money.
b What can you conclude about the value of y?

D

A02
A03

2.4 **Calculating the mean**

Objective

- You can calculate the mean for a set of data.

Why do this?

A swimmer could calculate the mean of their race times to see how well they have done throughout the year.

Get Ready

Find the sum of each of the following sets of numbers.
a 8   7   3   5   9   12
b 1.6   2.4   8.1   3.1   5.2

Key Points

- The mean of a set of data is the sum of the values divided by the total number of observations.
  This can be shown as the following formula

$$\text{mean} = \frac{\text{sum of values}}{\text{number of values}}$$

- For a sample of n values of x, the mean $= \frac{\sum x}{n}$.

  $\sum$ is the Greek letter s (sigma) which is short for sum. This, therefore, means the sum of the xs divided by n.

Example 16 | Find the mean of the numbers
2, 4, 6, 8, 10, 9, 6, 7.

$2 + 4 + 6 + 8 + 10 + 9 + 6 + 7 = 52$ ← Total the values.

The mean $= \frac{52}{8} = 6.5$.   ← Divide by the number of values to get the mean.

Example 17 | Over a career lasting 20 seasons, a footballer made a mean number of appearances for the English football team of 1.8 per season.
Work out how many times he played for England.

Total number of appearances $= 20 \times 1.8 = 36$.   ← Sum of the values = mean × number of values.

Example 18

The times, to the nearest tenth of a second, run by an athlete in his last ten 400-metre races were: 48.3, 47.2, 49.3, 50.4, 48.6, 60.0, 48.0, 48.2, 51.2, 47.2.

a Find the mode.

b Find the median.

c Find the mean.

a The mode is 47.2 seconds. ← 47.2 occurs twice; each of the other times occurs once.

b 47.2 47.2 48.0 48.2 48.3 48.6 49.3 50.4 51.2 60.0 ← There are two middle values: 48.3 and 48.6.

The median is 48.45 seconds. ← Halfway between is 48.45 seconds.

c The total of all values = 47.2 + 47.2 + 48.0 + 48.2 + 48.3 + 48.6 + 49.3 + 50.4 + 51.2 + 60.0 = 498.4 ← Find the total sum.

The mean = $\frac{498.4}{10}$ = 49.84 seconds. ← Divide by the total number of values.

Exercise 2D

1 A taxi driver has 11 calls during one day.
The numbers of passengers she carries on each journey are as follows.

5 4 1 3 6 4 3 1 2 1 3

Work out the mean number of passengers she carried per journey.

2 An academy has the following numbers of boys and girls in each year.

Year	Boys	Girls
7	134	128
8	138	130
9	160	141
10	162	154
11	156	150
12	110	125
13	92	110

a Find the mean number of boys per year.
b Find the mean number of girls per year.

3   During four weeks in July a man earns a mean wage of £323 per week.
     Work out how much he earns in total over the four-week period.

4   In a cricket match the eleven players scored the following numbers of runs.

     60   23   10   0   12   56   17   10   21   35   20

     a   Find the mode.
     b   Find the median number of runs.
     c   Find the mean number of runs.

2.5 Using the three types of average

◎ Objective

- You can discuss the advantages and disadvantages of the different measures of average.

◈ Why do this?

You would use the mode to work out the most popular meal in the school canteen, but you would use the median or mean to work out an average of how many people eat in the canteen each week.

◈ Get Ready

24   20   28   25   10   19   20

a   Work out the median and mode of these data.
b   Calculate the mean correct to 2 significant figures.

◈ Key Points

- Each of the three measures of average is useful in different situations.
- The following table will help you decide which of the three averages works best in different situations by showing a summary of the advantages and disadvantages of each measure.

Measure	Advantages	Disadvantages
MODE Use the mode when the data are non-numeric or when asked to choose the most popular item.	Extreme values (outliers) do not affect the mode. Can be used with qualitative data.	There may be more than one mode. There may not be a mode, particularly if the data set is small.
MEDIAN Use the median to describe the middle of a set of data that does have an extreme value.	Not influenced by extreme values.	Not as popular as mean. Actual value may not exist.
MEAN Use the mean to describe the middle of a set of data that does not have an extreme value.	Is the most popular measure. Can be used for further calculations. Uses all the data.	Affected by extreme values.

Example 19

A company consists of six workers, and their supervisor. The rates of pay of the six workers are £7, £7, £8, £9, £11 and £12 per hour. The supervisor is paid £25 per hour.

a Find the mode, median and mean rate of pay.

b Write down, giving a reason, which of the three averages you would use in the following situations.

   i When asked the typical wage rate.

   ii When trying to persuade a prospective employee to join the company.

a Mode = £7 per hour ⟵ | There are two £7 values.

Median = £9 per hour ⟵ | There are three values higher than £9 and three lower.

Total = 7 + 7 + 8 + 9 + 11 + 12 + 25 = 79 ⟵ | Total the values and divide by the total sum.

Mean = $\frac{79}{7}$ = £11.29

b **i** The mode of £7 is the lowest value partly because the number in the sample is small. You therefore would not use this as a 'typical' value. The median is probably the best measure in this case as it is unaffected by the high wage rate of the supervisor.

   ii There are only two values greater than the mean as the high wage rate of the supervisor has pulled the mean value up. As the mean is the highest average it is the one you would use to persuade a prospective employee to join the company.

ResultsPlus
Examiner's Tip

When comparing averages, look at how well each average represents the numbers as a whole, and give reasons why they would or would not be representative.

Exercise 2E

1 Five friends each buy a new dress for a party. They spend the following amounts of money.

   £17   £148   £22   £17   £31

   a Work out the mean, the mode and the median values.

   b Which average would best describe the amount of money they spent? Give a reason for your answer.

2 Write down one advantage and one disadvantage of using the mean as an average.

3 A restaurant records the number of diners it has every day for a week. The numbers are as follows.

   28   40   28   38   110   170   33

   a Write down the mode.

   b Work out the median number of diners.

   c Work out the mean number of diners.

   d The manager wishes to sell the restaurant. What average is he likely to use when talking to prospective buyers? Give a reason for your answer.

A03

D

2.6 Using calculators

Objectives

- You use the calculator effectively and efficiently.
- You know how to enter complex calculations and use the function keys.
- You make sensible estimates of a range of measures.

Why do this?

You are less likely to make numerical mistakes when using a calculator, and many functions are difficult to do longhand.

By estimating the answer you can check any calculations for accuracy

Get Ready

Using BIDMAS, work out

a $3^2 \times (2 + 7)$ **b** $\dfrac{4^2 \times 3}{(5 + 3)}$

Using a calculator

Key Points

- The calculator display is limited to a certain number of figures. Where possible, avoid rounding until supplying a final answer. You can use the calculator's memory to help with more complicated numbers.
- Scientific calculators usually have statistical functions, but their usage will vary between calculators. The calculator's instructions will explain how to enter and process statistical data.
- The negative sign for **directed numbers** looks very much like the subtraction sign.
- Basic calculators have a change sign key $\boxed{+/-}$. To enter a number such as -6, press $\boxed{6}$ then press $\boxed{+/-}$.
- Scientific calculators have the negative sign $\boxed{(-)}$. To enter a number such as -6, press $\boxed{(-)}$ then $\boxed{6}$.

Example 20   Work out $\dfrac{(-6.5) + 2}{2}$

On a basic calculator:

$\boxed{6}\ \boxed{\cdot}\ \boxed{5}\ \boxed{+/-}\ \boxed{+}\ \boxed{2}\ \boxed{\div}\ \boxed{2}\ \boxed{=}$ ← The display shows -2.25

On a scientific calculator:

$\boxed{(}\ \boxed{(-)}\ \boxed{6}\ \boxed{\cdot}\ \boxed{5}\ \boxed{+}\ \boxed{2}\ \boxed{)}\ \boxed{\div}\ \boxed{2}\ \boxed{=}$ ← The display shows -2.25

Finding roots and powers on a calculator

Key Points

- To work out **square numbers** use the $\boxed{x^2}$ button. (You may have to use the $\boxed{INV}$ key to get this.)
- To work out **cube numbers** some calculators have an $\boxed{x^3}$ button.
- To find a square root use the $\boxed{\sqrt{}}$ button.
- Scientific calculators have a power or index key $\boxed{x^y}$.
- There is also a root key $\boxed{x^{\frac{1}{y}}}$. (You may have to use the $\boxed{INV}$ key to get this.)

Example 21   Find

   a 6.3^2 b $\sqrt{9.61}$ c 2.3^4 d $\sqrt[4]{81}$.

a $\boxed{6}\,\boxed{\cdot}\,\boxed{3}\,\boxed{x^2}\,\boxed{=}$ ⟵   The display shows 39.69.

b $\boxed{\sqrt{\ }}\,\boxed{9}\,\boxed{\cdot}\,\boxed{6}\,\boxed{1}\,\boxed{=}$ ⟵   The display shows 3.1.

c $\boxed{2}\,\boxed{\cdot}\,\boxed{3}\,\boxed{x^y}\,\boxed{4}\,\boxed{=}$ ⟵   The display shows 27.9841.

d $\boxed{8}\,\boxed{1}\,\boxed{\text{INV}}\,\boxed{x^{\frac{1}{y}}}\,\boxed{4}\,\boxed{=}$ ⟵   The display shows 3.

Estimating

Key Point

◉ When doing a long calculation with decimals you can check your answer by calculating an **estimate** for the value.

Example 22   Estimate the value of $\dfrac{\sqrt{8.8}+3.4}{4.3}$.

This is approximately $\dfrac{\sqrt{9}+3}{4} = \dfrac{3+3}{4} = \dfrac{6}{4} = 1.5$ ⟵   The actual value is 1.5916198.

Exercise 2F

Use a calculator to complete questions 2 and 3.

1   Mia and her friends are going to a fancy dress party. She wants to know how much money her friends will have spent altogether on clothes for the party.
   Here is a list of what each person spent.
      £12.65, £18.25, £2.60, £22.40, £14.20, £12.70

   a   How much did they spend altogether?

   b   If they had still spent this total amount but each had spent the same amount of money, how much would each have spent?

A02
A03

2   Find the values of:
   a 4.6^2 b 8.9^3 c $\sqrt{2079.36}$
   d $\sqrt[3]{195112}$ e $7+(3\times6)-(9-4)$ f $-8\times(6-2)\times-3$

D

3   a   Estimate the value of $\dfrac{4.1^2\times3}{6.1}$

   b   Find the accurate value of $\dfrac{4.1^2\times3}{6.1}$ to 3 decimal places.

2.7 Using frequency tables to find averages

◎ Objective

● You can use a frequency table to find averages.

⑦ Why do this?

You can group some data in a frequency table, for example, goals scored in a football season or number of fans in the crowd per match.

⊕ Get Ready

The following numbers came up when a dice was thrown.

1 5 3 2 4 3 2 4
5 3 2 6 5 2 3 4

Represent these data as a frequency table.

Key Points

● When data are given in the form of a **frequency table**, the mode is the number that has the highest frequency.
● The median is the number that is the middle value, or halfway between the two middle values.
● To work out the mean for discrete data in a frequency table, use the following formula.

$$\text{Mean} = \frac{\sum f \times x}{\sum f}$$ where f is the frequency, x is the variable and $\sum$ means 'the sum of'.

Example 23

The following table shows information about the number of goals scored per match over two seasons by a football team.

Number of goals	0	1	2	3	4	5
Frequency	8	15	12	7	3	1

a Write down the mode of these data.
b Find the median of these data.
c Work out the mean of these data.

a The mode is 1 goal. ⟵ 1 has the highest frequency (15).

b

Number of goals (x)	Frequency (f)	Frequency × number of goals (f × x)
0	8	0
1	15	15
2	12	24
3	7	21
4	3	12
5	1	5
Total	46	77

There are 12 occasions when they scored 2 goals so there are 12 × 2 = 24 goals.

The total number of matches is 46.

The total number of goals is the sum of all the $f \times x$ values.

The total frequency is 46 so the median will be the 23.5th value. ⟵ $\dfrac{46+1}{2} = 23.5$
The median will be midway between the 23rd and 24th values.

There were 8 games with no goals scored.
There are $15 + 8 = 23$ games with 0 or 1 goals scored.
The 23rd value must be 1 and the 24th value must be 2.
The median is therefore $1\frac{1}{2}$.

c The mean is $\dfrac{77}{46} = 1.67$ goals. ⟵ $\text{Mean} = \dfrac{\text{Total number of goals}}{\text{Total frequency}} = \dfrac{\Sigma f \times x}{\Sigma f}$

Exercise 2G

1 The following table shows information about the numbers of siblings (brothers and sisters) a group of children have.

Number of siblings (x)	Frequency (f)	Frequency × number of siblings (f × x)
0	3	
1	8	
2	9	
3	4	
4	3	
5	0	
6	2	
7	1	
Total		

a Copy and complete the table.
b Write down the mode of these data.
c Work out the median number of siblings.
d Work out the mean number of siblings.

2 The table shows information about the numbers of paper clips in each packet of a box containing 60 packets.

Number of paper clips (x)	101	102	103	104	105	106	107
Frequency (f)	6	4	8	20	15	2	5

a Write down the mode of these data.
b Work out the median number of paper clips.
c Work out the mean number of paper clips.

3 Calgom Engineering employs apprentices.
The table shows information about the ages of the apprentices they have in 2008.

Ages of apprentices (x years)	17	18	19	20	21	22
Frequency (f)	10	8	8	2	2	2

a Write down the mode of these data.
b Work out the median age.
c Work out the mean age.

D

2.8 **Modal class and median of grouped data**

◉ **Objectives**

● You can find the modal class for grouped data.
● You can find a class interval containing the median of grouped data.

❓ **Why do this?**

In order to understand large amounts of data, you can put it into groups. This can be useful when looking at the speeds of cars on a motorway, or the number of albums sold in the UK.

⬆ **Get Ready**

Here are 30 numbers.

1 2 5 16 1 3 18 14 7 12
12 9 4 16 18 10 6 13 2 18
19 4 6 3 11 11 2 19 4 3

Draw up a frequency table for these data using the following class intervals.

1–4   5–9   10–14   15–19

🕐 **Key Points**

◉ When dealing with continuous data and class intervals, the class interval with the highest frequency is called the **modal class**.

◉ You cannot find the exact value of the median.
You can only write down the class interval in which the median falls.

🔍 **Example 24**

The frequency table below gives information about the number of phone calls made during a day by 21 people.

Number of calls (class interval)	Frequency
3−5	2
6−8	3
9−11	5
12−14	7
15−17	4

These are discrete data. Each whole number appears in only one class. There will be 3 + 2 = 5 values less than or equal to 8.

a Find the modal class.
b Find the class into which the median falls.

a The modal class is 12–14.   *Look for the class with the highest frequency.*

b There are 3 + 2 = 5 people that made less than 8 phone calls and 5 + 5 = 10 people that made less than 12 calls so the median is in the class interval 9–11 calls.   *The median will be the 11th value.*

Example 25 The following frequency table gives information about the speed (s), in miles per hour, of 50 cars.

Speed (s mph)	Frequency (f)
$20 \leqslant s < 25$	4
$25 \leqslant s < 30$	10
$30 \leqslant s < 35$	12
$35 \leqslant s < 40$	15
$40 \leqslant s < 45$	9

These are continuous data.

There will be 4 + 10 = 14 values less than 30.

a Find the modal class.

b Find the class into which the median falls.

a The modal class is $35 \leqslant s < 40$.

b The median falls in the class $30 \leqslant s < 35$.

There are 14 cars doing less than 30 mph and 26 doing less than 35 mph.

Exercise 2H

1 The weekly wages of the employees in a vehicle repair workshop are shown in the following grouped frequency table.

Weekly wage (£s)	Frequency (f)
£240–£280	4
£281–£320	20
£321–£360	12
£361–£400	14

a Write down the modal class.
b Find the class into which the median falls.

2 The table shows the weights of silver deposited on an electrode over 30 different experiments.

Weight (x g)	Frequency (f)
$0.30 \leqslant x < 0.35$	3
$0.35 \leqslant x < 0.40$	7
$0.40 \leqslant x < 0.45$	6
$0.45 \leqslant x < 0.50$	14

a Write down the modal class.
b Find the class into which the median falls.

D

modal class

D

3   A manufacturer produces steel machine parts.
     The lengths of a sample of 200 parts are shown in the table below.

Length (x cm)	Frequency (f)
$69.5 \leqslant x < 69.6$	2
$69.6 \leqslant x < 69.7$	10
$69.7 \leqslant x < 69.8$	30
$69.8 \leqslant x < 69.9$	34
$69.9 \leqslant x < 70.0$	35
$70.0 \leqslant x < 70.1$	56
$70.1 \leqslant x < 70.2$	33

a   Write down the modal class.
b   Find the class into which the median falls.

2.9 Estimating the mean of grouped data

⊙ Objective

○ You can estimate the mean
of grouped data.

⊘ Why do this?

You could work out the mean amount of time you spend on your mobile
phone each month in order to choose the best monthly tariff for you.

⊕ Get Ready

Which number is halfway between:

a   56 and 64   b   0.75 and 0.85   c   0.001 and 0.0001

Key Point

◉ An estimate for the mean of grouped data can be found by using the middle value of the class interval.

Example 26   Work out an estimate for the mean number of phone calls in Example 24.

Number of calls	Frequency (f)	Class mid point (x)	$f \times x$
3−5	2	4	8
6−8	3	7	21
9−11	5	10	50
12−14	7	13	91
15−17	4	16	64
Totals	21		234

The middle value of the class 3–5 is 4.
The middle value of the class 6–8 is 7.
The 3 people in the class 6–8 might
not all have made 7 calls.
This is why it is an estimated mean.

You can now use the formula.

Estimated mean $= \dfrac{\sum f \times x}{\sum f} = \dfrac{234}{21} = 11.14$ calls.

Example 27 Work out an estimate for the mean speed of the cars in Example 25.

Speed (s mph)	Frequency (f)	Class mid point (x)	f × x
20 ≤ s < 25	4	22.5	90
25 ≤ s < 30	10	27.5	275
30 ≤ s < 35	12	32.5	390
35 ≤ s < 40	15	37.5	562.5
40 ≤ s < 45	9	42.5	382.5
Totals	50		1700

Estimated mean = $\dfrac{1700}{50}$ = 34 mph.

Exercise 2I

1   The following group frequency table shows the ages of members of an aerobics class.

Age range (years)	16–25	26–35	36–45	46–55	56–65	66–75
Frequency	4	10	12	4	8	2

Work out an estimate for the mean age of the members.

2   The manager of a supermarket recorded the length of time, in seconds, that customers had to wait in the checkout queue. The results are shown in the grouped frequency table below.

Waiting time (t seconds)	0 ≤ t < 100	100 ≤ t < 200	200 ≤ t < 300	300 ≤ t < 400
Frequency	10	46	20	8

Work out an estimate for the mean waiting time of the shoppers.

3   A call centre kept a record of the time, in seconds, that callers had to wait to speak to call centre staff over a period of 10 minutes. The results are shown in the grouped frequency table.

Waiting time (t seconds)	0 ≤ t < 30	30 ≤ t < 60	60 ≤ t < 90	90 ≤ t < 120	120 ≤ t < 150
Frequency	30	55	26	13	6

Work out an estimate for the mean waiting time of the callers.

C

2.10 Range, quartiles and interquartile range

- You can calculate the range of a set of data.
- You can work out the quartiles of a set of data.
- You can find the interquartile range for a set of data.
- You can compare data sets using a measure of average and a measure of range.

⟡ Why do this?

To get a better idea of the heights of students in your class, it helps to work out the range and quartiles.

⬆ Get Ready

1. Arrange the following numbers in ascending order.

   43 21 18 32 45 16 16 14 23 27 38 49

2. Arrange the following weights in ascending order.

   56.2 kg   43.4 kg   56.2 kg   49.9 kg   43.5 kg   36.0 kg

🔍 Key Points

- **Range** = highest value of a data set − lowest value of a data set
- The **quartiles**, Q_1, Q_2 and Q_3, split the data into four parts.

Lowest value	Q_1	Q_2	Q_3	Highest value
	25% of data	25% of data	25% of data	25% of data

- For a set of data arranged in **ascending order**:
  - Q_1, the **lower quartile**, is a quarter of the way through the data
  - Q_2, the second quartile, is halfway through the data (the median)
  - Q_3, the **upper quartile**, is three-quarters of the way through the data.
- Generally for a set of n data values arranged in ascending order:
  - Q_1 is the $\left(\dfrac{n+1}{4}\right)$th value
  - Q_2 (median) is the $\left(\dfrac{n+1}{2}\right)$th value
  - Q_3 is the $\left(\dfrac{3(n+1)}{4}\right)$th value.
- The **interquartile range** (IQR) = $Q_3 - Q_1$.
- The average and the range together give a description of the **distribution** of the data.
- To compare the distributions of sets of data you need to give a measure of average and a measure of spread.

🔍 Example 28

A03

Eight students sat two examinations. Their marks, out of 30, are shown below.

Mathematics: 20, 16, 30, 17, 25, 21, 22, 19.

English: 14, 16, 23, 28, 24, 12, 21, 13.

a Work out the range of marks for each exam.

b Which set of marks was the more consistent? Give a reason for your answer.

a Range for mathematics exam = 30 − 16 = 14.
  Range for English exam = 28 − 12 = 16.
b The mathematics marks were more consistent.
  This is because the range was smaller.

Example 29 Find the quartiles and interquartile range of the following set of data:
570, 460, 600, 480, 500, 510, 340, 560, 320, 590, 650.

320 340 **460** 480 500 **510** 560 570 **590** 600 650 ⟵ | Write the data in order, starting with the lowest value.

$Q_1 = 460$ $Q_2 = 510$ $Q_3 = 590$ ⟵ | Find the quartiles.

IQR = 590 − 460 = 130

Example 30 The following data give information about the heights, in metres, of trees commonly found in English hedgerows: 20, 18, 30, 10, 31, 4, 12, 18, 27, 7, 24, 24, 30, 6, 10.
Find the upper and lower quartiles and interquartile range of the heights.

4 6 7 10 10 12 18 18 20 24 24 27 30 30 31 ⟵ | Put the data in ascending order.

$n = 15$

$Q_1 = \dfrac{15 + 1}{4} = $ 4th value = 10 m ⟵ | Use the formula $\dfrac{n+1}{4}$.

$Q_3 = \dfrac{3(15 + 1)}{4} = $ 12th value = 27 m ⟵ | Use the formula $\dfrac{3(n+1)}{4}$.

IQR = 27 − 10 = 17 m ⟵ | IQR = $Q_3 − Q_1$

Example 31 The heights, in cm, of 11 men and their sons are given below.
Men's heights: 150, 152, 155, 160, 165, 170, 175, 180, 180, 190, 198.
Sons' heights: 163, 166, 168, 170, 170, 173, 175, 178, 183, 183, 185.
  a Find the means and interquartile ranges of these data.
  b Compare and contrast these results.

A03

a Men's heights:

Mean = $\dfrac{150 + 152 + 155 + 160 + 165 + 170 + 175 + 180 + 180 + 190 + 198}{11}$

   = 170.45 cm

$Q_1 = (\dfrac{11 + 1}{4} = $ 3rd value) = 155 cm

$Q_3 = \dfrac{3(11 + 1)}{4} = $ 9th value = 180 cm ⟵ | The data are already ordered. Use the formula to find the quartiles.

IQR = 180 − 155 = 25 cm

Sons' heights:

Mean = $\dfrac{163 + 166 + 168 + 170 + 170 + 173 + 175 + 178 + 183 + 183 + 185}{11}$

   = 174 cm

$Q_1 = 168$ cm, $Q_3 = 183$ cm, IQR = 183 − 168 = 15 cm | Mean = $\dfrac{\text{total of values}}{\text{total frequency}}$

b The mean height of the sons was higher than that of their fathers.
Sons are generally taller than their fathers.
The IQR of the fathers was higher than that of their sons.
The fathers' heights were more spread out.

| The mean is used in preference to the median as there are no extreme values.

Exercise 2J

B

1 Write down another name for Q_2.

2 Eleven college students were asked to record the amount of time they spent on the internet one evening. Their times, in minutes, were:

38   42   50   56   60   62   65   70   70   75   80

   a Write down Q_1, Q_2 and Q_3 for these data.

   b Work out the interquartile range.

   c Work out the range.

3 A lepidopterist set a moth trap for 15 evenings. She recorded the number of moths trapped. They were:

  5   9   15   12   21   14   19   8

11   24   16   13   20   7   6

   a Write down Q_1, Q_2 and Q_3 for these data.

   b Work out the interquartile range.

   c Work out the range.

4 The number of bags of crisps sold per day in a general shop was recorded over 13 days. The results are shown below.

32   45   36   56   45   68   29   48

21   45   32   47   59

   a Write down Q_1, Q_2 and Q_3 for these data.

   b Work out the interquartile range.

   c Work out the range.

Chapter review

- The **mode** of a set of discrete data is the value that occurs most frequently.
- The **median** is the middle value when the data are ordered from the smallest to the largest.
- If there are two middle values in a set of data, the median is halfway between them.
- The **mean** of a set of data is the sum of the values divided by the total number of observations.
- For a sample of n values of x, **mean** $= \dfrac{\textbf{sum of values}}{\textbf{number of values}} = \dfrac{\sum x}{n}$
- For discrete data in a frequency table,

  mean $= \dfrac{\sum f \times x}{\sum f}$ where f is the frequency, x is the variable and $\sum$ means 'the sum of'.
- For grouped data:
  - The class interval with the highest frequency is called the **modal class**.
  - You can only write down the class interval in which the median falls.

- An estimate for the mean of grouped data can be found by using the middle value of the class interval.
- **Range** = highest value − lowest value.
- For a set of data arranged in ascending order, the **quartiles**, Q_1, Q_2 and Q_3, split the data into four parts:
  - Q_1, the lower quartile, is a quarter of the way through the data
  - Q_2, the second quartile, is halfway through the data
  - Q_3, the upper quartile, is three-quarters of the way through the data.
- Generally for a set of n data values arranged in ascending order:
  - Q_1 is the $\left(\dfrac{n+1}{4}\right)$th value
  - Median Q_2 is the $\left(\dfrac{n+1}{2}\right)$th value
  - Q_3 is the $\left(\dfrac{3(n+1)}{4}\right)$th value.
- The **interquartile range** (IQR) = $Q_3 - Q_1$.
- To compare the distributions of sets of data you need to give a measure of average and a measure of spread.

Review exercise

1. Peter rolled a 6-sided dice ten times.
   Here are his scores.

   | 3 | 2 | 4 | 6 | 3 | 3 | 4 | 2 | 5 | 4 |

   a Work out the median of his scores.

   b Work out the mean of his scores.

   c Work out the range of his scores. *June 2007*

2. Tom and Jessica decide to make cakes for the village party.
   Jessica is going to make a batch of shortbread biscuits.
   She mixes 165 g of butter, 75 g sugar and 260 g plain flour.
   Each biscuit uses 24 g of mixture.
   a Work out the largest number of biscuits Jessica can make.

   Tom decides to make a batch of oatcake.
   He mixes 100 g jumbo oats, 120 g porridge oats, 160 g brown sugar and 220 g margarine.
   He makes 24 oatcakes all the same size and has no mixture left.
   b How much mixture did he use for each oatcake?

   A02

3. Nine friends go to a charity shop. They spend the following amounts of money:

   | £4 | £6 | £4 | £38 | £10 | £4 | £3 | £7 | £5 |

   a Work out the mode, the median and the mean of the amounts they spent.

   b Which of these three averages best describes the amount they spent? Give a reason for your answer.

   D

4. Write down one advantage and one disadvantage of using each of the following as an average.

   a The mode

   b The median

   c The mean

D

5   The frequency table below shows the number of aeroplanes that took off from a small airport each hour during one day in January 2010.

Number of aeroplanes	Frequency
0	3
1	1
2	2
3	2
4	8
5	5
6	3

a   Work out how many aeroplanes took off in total during the day.
b   Work out the mean number of aeroplanes taking off per hour.

6   Ali found out the number of rooms in each of 40 houses in a town.
He used the information to complete the frequency table.

Number of rooms	Frequency	
4	4	
5	7	
6	10	
7	12	
8	5	
9	2	

Ali said that the mode is 9.
Ali is wrong.

a   Explain why.
b   Calculate the mean number of rooms.

Nov 2007

7   Majid carried out a survey of the number of school dinners 32 students had in one week.
The table shows this information.

Number of school dinners	0	1	2	3	4	5
Frequency	0	8	12	6	4	2

Calculate the mean.

Nov 2008

8   The mean of eight numbers is 41.
The mean of two of the numbers is 29.
What is the mean of the other six numbers?

ResultsPlus
Exam Question Report

87% of students answered this sort of question poorly.

June 2007

9   A man kept a record of the number (x) of junk emails he received each day over a period of 100 days.
Given $\Sigma x = 770$, work out the mean number of junk emails he received each day.

10 The wing spans of 30 Emperor Dragonflies were measured.
The results are shown in the following grouped frequency table.

Wing span (l cm)	Frequency (f)
9.6 – 9.8	3
9.9 – 10.1	4
10.2 – 10.4	9
10.5 – 10.7	14

a Find the modal class.

b Find the class into which the median falls.

11 A vet keeps a record of the weight of all dogs brought into his surgery.
The table shows the adult weights (w kg) of the labradors in his records.

Class interval	Frequency (f)	Class mid point	$f \times x$
$26 \leqslant w < 29$	4		
$29 \leqslant w < 32$	7		
$32 \leqslant w < 35$	15		
$35 \leqslant w < 38$	12		
$38 \leqslant w < 41$	2		
Totals			

a Copy and complete the table.

b Work out an estimate for the mean weight of the dogs.

12 A researcher was conducting a study into the growth patterns of mice. She recorded the body length of 28 mice. The lengths (l mm) are shown in the following grouped frequency table.

Class interval	Frequency (f)
$70 \leqslant x < 75$	3
$75 \leqslant x < 80$	3
$80 \leqslant x < 85$	5
$85 \leqslant x < 90$	12
$90 \leqslant x < 95$	5

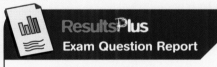

ResultsPlus
Exam Question Report

81% of students answered this sort of question poorly.

Work out an estimate for the mean body length of the mice.

13 Josh asked 30 students how many minutes they each took to get to school.
The table shows some information about his results.

Time (t minutes)	Frequency
$0 < t \leqslant 10$	6
$10 < t \leqslant 20$	11
$20 < t \leqslant 30$	8
$30 < t \leqslant 40$	5

Work out the estimate for the mean number of minutes taken by the 30 students. *Nov 2008*

14 The table gives information about the times, in minutes, that 106 shoppers spent in a supermarket.

Time (t minutes)	Frequency
$0 < t \leqslant 10$	20
$10 < t \leqslant 20$	17
$20 < t \leqslant 30$	12
$30 < t \leqslant 40$	32
$40 < t \leqslant 50$	25

a Find the class interval that contains the median.

b Calculate an estimate for the mean time that the shoppers spent in the supermarket.
Give your answer correct to 3 significant figures.

Nov 2007

15 Sethina recorded the times, in minutes, taken to repair 80 car tyres.
Information about these times is shown in the table.

Time (t minutes)	Frequency
$0 < t \leqslant 6$	15
$6 < t \leqslant 12$	25
$12 < t \leqslant 18$	20
$18 < t \leqslant 24$	12
$24 < t \leqslant 30$	8

Calculate an estimate for the mean time taken to repair each car tyre.

June 2009

* 16 There are 50 students in each of the year groups at a school.
A survey was carried out to find how many pets these students owned.
The table shows these results.

Number of pets	0	1	2	3	4
Year 9	1	29	14	5	1
Year 10	5	22	19	4	0
Year 11	32	11	6	1	0

Which year group has the least number of pets?
You must show all your calculations.

17 As part of an ongoing research programme, the pups in a small colony of grey seals were weighed, to the nearest kilogram, at the age of four weeks. Their weights were as follows:

42   40   45   47   50   48   39   47   42   50   49

a Write down Q_1, Q_2 and Q_3 for these data.

b Work out the interquartile range.

c Work out the range of the weights.

18 A council is introducing a new traffic management scheme to speed up morning rush-hour traffic. Before the scheme, 11 council workers are asked to record their journey time to work one Wednesday morning. After the scheme was put in place the same 11 workers were asked to record their journey time again one Wednesday morning. The results are shown in the table below.

Worker	A	B	C	D	E	F	G	H	I	J	K
Before (min)	23	30	10	13	15	22	16	19	21	14	15
After (min)	20	25	8	13	12	16	14	17	19	10	11

a Find the mean time taken before the traffic scheme was introduced.

b Find the mean time taken after its introduction.

c Find Q_1 and Q_3 for both sets of data.

d Find the interquartile ranges.

e Compare the time taken before the introduction of the traffic scheme with the time taken after it was introduced.

* 19 Ten people work in a small factory. The table shows their salaries.

Employees	Salary
1 owner	£180 000
1 manager	£40 000
8 workers	£10 000

The workers want a pay rise, but the owner doesn't want to give them a rise.

Explain how both the owner and the workers could use the word 'average' to justify their case.

* 20 Explain the following sentence:

The vast majority of dogs in this country have more than the average number of legs.

3 PROCESSING, REPRESENTING AND INTERPRETING DATA

When you buy food, the packaging gives you information about the nutritional value of that food but you will need to interpret it to understand what it means for your health. For example, a grilled salmon fillet gives you 30g of protein but unless you know that a woman needs approximately 46g of protein a day and a man approximately 56g, this is of little use. Now you can work out that for a woman, the salmon fillet gives her about 65% of her daily protein intake and for a man about 54%.

◎ Objectives

In this chapter you will be able to produce and interpret the following, for various types of data:

- ◉ pie charts
- ◉ stem and leaf diagrams
- ◉ bar charts and composite bar charts
- ◉ frequency diagrams
- ◉ histograms for continuous data
- ◉ frequency polygons
- ◉ cumulative frequency graphs
- ◉ box plots.

◆ Before you start

You need to be able to:

- ◉ measure and draw angles to the nearest degree
- ◉ measure and draw lines to the nearest mm
- ◉ understand grouped data.

3.1 Producing pie charts

Objectives

○ You can represent the proportions of different categories of data using a pie chart.

○ You can use the properties of angles at a point.

Why do this?

When a council collects council tax they like to show the taxpayers how they are spending their money. They might use a pie chart to show the proportions spent on different things.

Get Ready

1. How many degrees are there in a circle?

2. How many degrees are there in a quarter-circle?

3. What is **a** $\frac{1}{3}$ of 360   **b** $\frac{2}{8}$ of 180   **c** $\frac{4}{6}$ of 90?

Properties of angles

Key Points

◉ In a full turn there are 360°.

◉ In a quarter turn there are 90°.
A **right angle** is often marked on a diagram with a small square.

◉ In a half turn there are 180°.

◉ In three quarters of a turn there are 270°.

◉ The compass **bearing** of a point is measured in this way, with North being at 0°, East at 90°, South at 180° and West at 270°.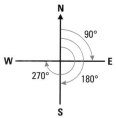

◉ Angles a, b and c are known as **angles at a point**.
If you add together angles a, b and c you get a full turn.
There are 360° in a full turn.
Angles at a point add up to 360°.

Example 1

Work out the size of angle a.

$110° + 142° = 252°$ ← Add 110° and 142° together.

$360° - 252° = 108°$

$a = 108°$ ← Subtract the result from 360°.

Exercise 3A

Questions in this chapter are targeted at the grades indicated.

1 Work out the size of angle b.

2 The big hand of a clock moves round from 3 to 5.
   a **i** Write down the size of the angle it has gone through.   **ii** How many minutes does this represent?
   The hour hand of the clock moves through 75°.
   b Write down the number of hours that have passed.

3 A pie chart is divided into 5 sectors. Four of the angles are 70°, 45°, 30°, and 55°.
   Write down the angle of the fifth sector.

Pie Charts

Key Points

○ A **pie chart** is often used to show data. It shows how the total is split up between the different categories.
○ In a pie chart the area of the whole circle represents the total number of items.
○ The area of a **sector** represents the number of items in the category represented by that sector.
   This pie chart shows how the population of the United Kingdom is split between
   the different countries.
   It shows that the least number of people live in Northern Ireland.
   The greatest number of people live in England.

○ The angles at the centre must add up to 360°.
○ The angle for a particular sector is found as follows: sector angle $= \dfrac{\text{frequency} \times 360°}{\text{total frequency}}$

Example 2

The table shows the number of theatre-goers who attended each type of performance at
least once in a 1-year period.

Performance	Musical	Play	Entertainment	Dance	Opera
Number	38	27	14	17	24

Draw a pie chart to represent this information.

Musical $\dfrac{38}{120} \times 360° = 114°$

> Total frequency = 38 + 27 + 14 + 17 + 24 = 120
> Use the formula to find each angle.

Play $\dfrac{27}{120} \times 360° = 81°$

Entertainment $\dfrac{14}{120} \times 360° = 42°$

Dance $\dfrac{17}{120} \times 360° = 51°$

Opera $\dfrac{24}{120} \times 360° = 72°$

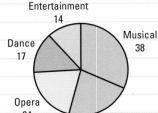

ResultsPlus
Examiner's Tip

Add the angles together to
make sure they add up to 360°.

Check: 114 + 81 + 42 + 51 + 72 = 360

Exercise 3B

1 The numbers of drinks dispensed by a vending machine in one day are shown in the table.
A pie chart is to be drawn to illustrate these data.

Type of drink	Tea	Black coffee	Chocolate	Orange	Coke	Latte
Number of drinks	54	42	18	30	12	24

Draw a pie chart to represent these data. Use a radius of 4 cm.

2 The snack bar at a bus station sold 120 sandwiches one lunch time.
The table shows the number of each type of sandwich sold.
A pie chart is to be drawn to illustrate these data.

Type of sandwich	Cheese	BLT	Tuna	Prawn	Ham	Chicken
Number of drinks	10	35	20	30	15	10

Draw a pie chart to represent these data. Use a radius of 4 cm.

3 A factory manager asks the employees how they travel to work.
The table shows these data. A pie chart is to be drawn to illustrate these data.

Method of getting to work	Walk	Car	Cycle	Train	Motorbike
Number of employees	14	32	12	10	4

Draw a pie chart to represent these data.

3.2 Interpreting pie charts

Objective

○ You can interpret a pie chart.

Why do this?

Election results are often shown in a pie chart.
You can interpret these graphs to see how many people voted for each party.

Key Points

◉ To read frequencies from a pie chart use the formula

$$\text{Frequency} = \frac{\text{sector angle} \times \text{total frequency}}{360°}$$

Example 3 The pie chart shows the number of Bronze Age finds made with a metal detector and the outcomes when they were submitted to a local museum.

There were 36 finds altogether.

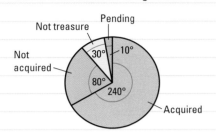

ResultsPlus

Watch Out!

In an exam 'work out' means calculate the frequency, so don't just measure the angle.

a Which type of outcome was most frequent?

b Work out the frequency for each outcome.

a 'Acquired' was the most common outcome.

b Acquired frequency = $\dfrac{\text{sector angle} \times \text{total frequency}}{360°} = \dfrac{240° \times 36}{360°} = 24$

ResultsPlus

Examiner's Tip

Always add up the frequencies for each sector to make sure they total to the right number.

Not acquired frequency = $\dfrac{80° \times 36}{360°} = 8$

Not treasure frequency = $\dfrac{30° \times 36}{360} = 3$

Pending frequency = $\dfrac{10° \times 36}{360} = 1$

Check: $24 + 8 + 3 + 1 = 36$

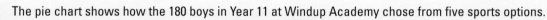

Exercise 3C

1 The pie chart shows how the 180 boys in Year 11 at Windup Academy chose from five sports options.

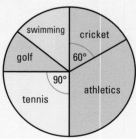

a Write down the least popular option.

b Write down the most popular option.

c Work out how many boys chose tennis.

d Work out how many boys chose cricket.

* **2** A company owns two coffee shops in Twyfield.
They do a survey to find the number of each type of coffee they sell between 9 am and 10 am on one particular day.

Coffee type	Frequency	
	Shop A	Shop B
Espresso	5	5
Americano	15	12
Latte	10	24
Mocha	40	20
Cappuccino	20	11

Compare and contrast the information by drawing two pie charts.

3.3 Representing and interpreting data in a stem and leaf diagram

Objectives

- You can represent data as a stem and leaf diagram.
- You can use a stem and leaf diagram to find the mode, median, range and quartiles of a set of data.

Why do this?

If you surveyed the number of DVDs that your friends have, you could use a stem and leaf diagram to show the pattern of the results.

Get Ready

Write the numbers in each set in order of size, with smallest number first.

a 65, 54, 72, 50   **b** 4.3, 4.6, 4.0, 4.4   **c** 0.11, 0.1, 0.01, 0.12

Key Points

- A **stem and leaf diagram** is a way of presenting data that makes it easy to see the pattern without losing the actual data.
- A stem and leaf diagram should always have a key.
- From a stem and leaf diagram you can find statistics about the data. The lower quartile (Q_1) is the value a quarter of the way through the data, the second quartile (Q_2) or median is halfway through, and the upper quartile (Q_3) is three-quarters of the way through.
- The interquartile range (IQR) is the difference between the upper and lower quartiles = $Q_3 - Q_1$.

Example 4

Here are the numbers of cigarettes smoked per day by 15 people who were attempting to give up smoking:

20, 35, 40, 42, 32, 15, 22, 30, 28, 34, 40, 43, 28, 41, 25

a Write these data as an **ordered stem and leaf diagram**.
b Write down the mode of these data.
c Find the median of these data.
d Work out the range of these data.
e Find the lower and upper quartiles and interquartile range.

a

Stem	Leaf				
1	5				
2	0	2	8	8	5
3	5	2	0	4	
4	0	2	0	3	1

Key 1|5 stands for 15

> The digit that each number begins with is called the stem.

> The following digit is called the leaf.

> Under stem, write the numbers 1 to 4.

> Opposite each stem, write the leaves. Don't worry about the order. This gives you an unordered stem and leaf diagram.

1	5				
2	0	2	5	8	8
3	0	2	4	5	
4	0	0	1	2	3

Key 1|5 stands for 15

> Next draw a stem and leaf with the leaves in order, starting with the smallest. This is an ordered stem and leaf diagram as asked for in the question.

b There are two modes: 28 and 40.

> Each appears twice, the others only once.

c The median is 32.

> 32 is the middle value.

d The range is $43 - 15 = 28$

> The range is the difference between the largest and smallest values. The largest and smallest values are the first leaf and the last leaf.

e $Q_1 = 25$ $Q_3 = 40$
  $IQR = 40 - 25 = 15$

> $Q_1 = \frac{16}{4}$th value = 4th value
> $Q_3 = 3 \times \frac{16}{4}$th value = 12th value
> You can find the values by counting in from each end.

Exercise 3D

1 Nassim records the number of emails he receives every day for 35 days.
The data he collects are shown in the stem and leaf diagram.

0	6	7	9	9					
1	4	7	7	8	8	9	9		
2	2	3	5	5	6	7	8	9	9	9
3	1	5	6	6	6	6	7		
4	3	6	8	9					
5	2	3	3						

Key 3 | 1 stands for 31

a Write down the mode of these data.

b Find the median of these data.

c Work out the range of these data.

d Find Q_1 and Q_3 of these data.

e Work out the interquartile range for these data.

2 Here are the number of minutes a sample of 19 people had to wait to see a dentist.

10	12	8	9	21	24	17	4	28	30
5	7	9	15	7	9	14	9	6	

a Draw an ordered stem and leaf diagram for these data.

b Use your stem and leaf diagram to find the mode of these data.

c Use your stem and leaf diagram to find the median of these data.

d Work out the range of these data.

e Use your stem and leaf diagram to find Q_1 and Q_3 of these data.

f Work out the interquartile range for these data.

3 A delivery driver does a journey on 23 days every month.
Here are the distances, in kilometres, that he travelled in March.

56	74	83	74	65	92	52	59
64	68	72	94	82	63	74	65
88	69	68	85	68	74	63	

a Draw an ordered stem and leaf diagram for these data.

b Use your stem and leaf diagram to find the mode of these data.

c Use your stem and leaf diagram to find the median of these data.

d Work out the range of these data.

e Use your stem and leaf diagram to find Q_1 and Q_3 of these data.

f Work out the interquartile range for these data.

3.4 Interpreting comparative and composite bar charts

⊙ **Objectives**

○ You can interpret comparative bar charts.
○ You can interpret composite bar charts.

? **Why do this?**

You may want to compare the sales of various categories of music in two shops. Composite bar charts would allow you to do this.

◈ **Get Ready**

What can you say about the data in these two charts?

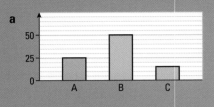
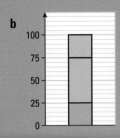

◉ **Key Points**

◉ Composite bar charts (sometimes called comparative or **component bar charts**) can be drawn to compare data. A dual bar is a type of comparative bar chart.

◉ In a comparative bar chart, two (or more) bars are drawn side-by-side for each category, and the heights of the bars can be compared category-by-category.

◉ A composite bar chart shows the size of individual categories split into their seperate parts.

✎ **Example 5**   The **dual bar chart** shows the number of houses sold by two agents in four months.

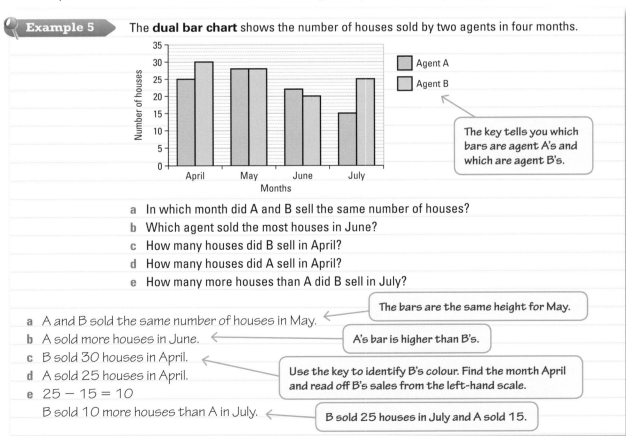

a In which month did A and B sell the same number of houses?
b Which agent sold the most houses in June?
c How many houses did B sell in April?
d How many houses did A sell in April?
e How many more houses than A did B sell in July?

a A and B sold the same number of houses in May. ← The bars are the same height for May.
b A sold more houses in June. ← A's bar is higher than B's.
c B sold 30 houses in April. ←
d A sold 25 houses in April. ← Use the key to identify B's colour. Find the month April and read off B's sales from the left-hand scale.
e 25 − 15 = 10
   B sold 10 more houses than A in July. ← B sold 25 houses in July and A sold 15.

Example 6 This composite bar chart shows the amounts of protein, carbohydrate, fat and fibre in 100 g of white and wholemeal flour.

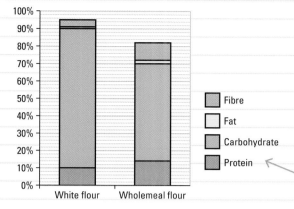

a How many grams of protein are there in 100 g of white flour?

b How many grams of carbohydrate are there in 100 g of wholemeal flour?

c Write down the flour which had the greater amount of fibre.

d Write down the flour with the smaller amount of fat.

e How many grams of wholemeal flour were not protein, carbohydrate, fat or fibre?

The key tells you the colour for each constituent.

a 10 g ← Identify the bar for white flour and use the colour key to find out which is protein.

b 70 − 14 = 56 g ← Read off from scale.

c Wholemeal ← The higher bar for fibre was for wholemeal flour (10 g).

d White ← The shorter bar for fat was for white flour (1 g).

e 100 − 82 = 18 g ← Read off the total height of the bar for wholemeal and take it away from 100 g (the figures are per 100 g).

Exercise 3E

1 The composite bar chart shows the temperature in a number of resorts in April and October.

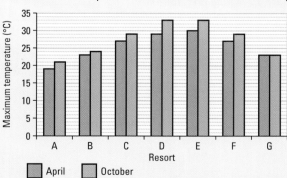

a Write down the maximum temperature in April.

b Write down the maximum temperature in October.

c Write down the resort that had the same maximum temperature in both months.

d Write down the resorts in which the maximum temperature in October was 29°C.

e Write down the resort in which the maximum temperature in April was 19°C.

2 The composite bar chart shows how David spends his money.

a What did David spend most on?

b What did David spend least on?

c What percentage of his income did he spend on housing?

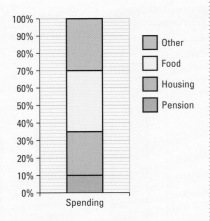

3 The composite bar charts show the make up of 100 grams of each of two cereals: Weeties and Fruitbix.

a How many grams of carbohydrate are there in 100 g of Weeties?

b Estimate the number of grams of fat in 100 g of Fruitbix.

c Write down the name of the cereal that has more fibre.

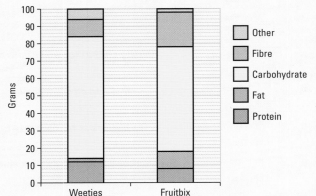

4 Market days in Ulvston are on Thursday and Saturday. Mattie runs a market stall that sells jumpers on each of these days. The composite bar chart shows his sales in one week in November.

a On which day were most jumpers sold overall?

b On which day were most green jumpers sold?

c How many red jumpers were sold on Saturday?

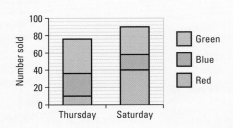

3.5 Drawing and interpreting frequency diagrams and histograms

Objectives

● You can draw a frequency diagram for grouped discrete data.

● You can draw a histogram for continuous data.

Why do this?

An exam board may choose to illustrate how the candidates in a particular year performed in one of its exams by creating a frequency diagram or histogram with the data.

Get Ready

What is the width of each class interval?

a $0 \leqslant h < 3$ b $8 \leqslant h < 24$ c $75 \leqslant h < 100$

Key Points

- A **frequency diagram** can be drawn from grouped discrete data.
- A frequency diagram for grouped discrete data looks the same as a bar chart except that the label underneath each bar represents a group.
- A **histogram** can be drawn from grouped continuous data.
- A histogram is similar to a bar chart but represents continuous data so there is no gap between the bars.

Example 7

The table shows the number of pizzas ordered in a restaurant from 7 pm to 8 pm on consecutive nights. Draw a frequency diagram for the information in the frequency table.

Number ordered	Frequency
1–5	2
6–10	4
11–15	8
16–20	6

There is a gap between the bars because, for example, there is no whole number between 15 and 16.

Example 8

The **grouped frequency table** shows information about the lengths of a series of roadworks.

a Write down the modal class interval.

b The length of one set of roadworks is 177.2 m. In which class interval is this recorded?

c The length of another set is exactly 180 m. In which class interval is this length recorded?

d Draw a histogram for these data.

Length (l metres)	Frequency
$160 \leqslant l < 165$	10
$165 \leqslant l < 170$	14
$170 \leqslant l < 175$	8
$175 \leqslant l < 180$	5
$180 \leqslant l < 185$	3
$185 \leqslant l < 190$	2

a The modal class is $165 \leqslant l < 170$.

This class interval has the highest frequency, 14.

b This set is in the class interval $175 \leqslant l < 180$.

177.2 m is greater than 175 m but less than 180 m.

c This set is in the class interval $180 \leqslant l < 185$.

180 is shown at the end of one class interval and at the beginning of another. The sign for 'less than or equal to' ($\leqslant$) shows that 180 m should go in the class interval $180 \leqslant l < 185$.

d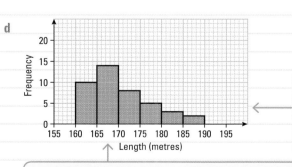

In this histogram the area of the bars is proportional to the frequency. In the class interval 160 to 165 there are 50 little squares representing a frequency of 10. Each little square is equal to a frequency of $\frac{1}{5}$.

In the class interval 165 to 170 there are 70 little squares so it represents a frequency of $70 \times \frac{1}{5} = 14$.

Exercise 3F

D

1   The grouped frequency table shows information about
the number of computer games owned by each of
35 college students.

Draw a frequency diagram for this information.

Number of games	Frequency
0 to 2	2
3 to 5	5
6 to 8	9
9 to 11	12
12 to 14	7

2   The grouped frequency table shows information about
the wingspans of 36 snowy owls.

a   Write down the modal class.

b   The first snowy owl measured had a wingspan of
140 cm. In which class interval is this recorded?

c   Draw a histogram for these data.

Wingspan (w cm)	Frequency
$125 \leqslant w < 130$	2
$130 \leqslant w < 135$	10
$135 \leqslant w < 140$	14
$140 \leqslant w < 145$	7
$145 \leqslant w < 150$	3

3   In a research project 40 young otters were weighed.
Some information about their weights is shown in the table.

a   Write down the modal class.

b   In which class interval does the weight of 137 g fall?

c   Draw a histogram for these data.

Weight (w g)	Frequency
$135 \leqslant w < 137$	3
$137 \leqslant w < 139$	10
$139 \leqslant w < 141$	14
$141 \leqslant w < 143$	8
$143 \leqslant w < 145$	5

3.6 **Drawing and using frequency polygons**

Objectives

- You can draw frequency polygons.
- You can recognise simple trends from a frequency
  polygon.
- You can use two polygons to make comparisons
  between two sets of data.

Why do this?

If you take a sample of your classmates' long-jump
results a frequency polygon would give you a good
idea of how the lengths are distributed.

Get Ready

Which number is halfway between:

a   3 and 7   b   15 and 20   c   112 and 119?

Key Points

◉ A **frequency polygon** is another graph which shows data.

◉ When drawing a frequency polygon you draw a histogram then mark the mid points of the tops of the bars and join these with straight lines.

◉ More than one frequency polygon can be drawn on the same grid to compare data.

Example 9   Draw a frequency polygon for the data in Example 8.

ResultsPlus
Watch Out!

Don't forget that the points are plotted at the mid points of the class intervals.

Example 10   The frequency table gives information about the time waited, in seconds, at a set of traffic lights.

a Write down the modal class.

b Use the information to draw a histogram.

c Draw a frequency polygon to represent the information.

Time waited (t seconds)	Frequency
$90 \leqslant t < 95$	6
$95 \leqslant t < 100$	6
$100 \leqslant t < 105$	7
$105 \leqslant t < 110$	4
$110 \leqslant t < 115$	5
$115 \leqslant t < 120$	2

a *The modal class is* $100 \leqslant t < 105$.

b, c

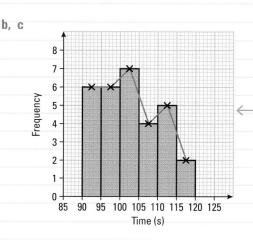

As the question asks for both a histogram and a frequency polygon to be drawn, draw the histogram first.

Example 11 These two frequency polygons show the heights of seedlings growing in two different composts.

Compare the heights of the two groups.

Give reasons for your answers.

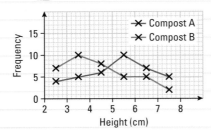

Compost A gives taller seedlings overall. ← Above 5 cm, the line showing the heights with compost A is above the line for compost B.

There are more very tall seedlings with compost A. ← There are five seedlings in the 7 − 8 cm class interval which were grown in compost A compared to two for compost B.

There are more very short seedlings with compost B. ← There are seven seedlings grown in compost B but only four for compost A in the 2 − 3 cm class interval.

Exercise 3G

1 A seed producer wants to know the numbers of peas in pods of a new variety of peas.
He records the number of peas in 60 pods. The table shows this information.

Number of peas	3	4	5	6	7	8
Frequency	2	4	7	10	22	15

Draw a frequency polygon for these data.

2 The noise levels at 40 locations near an airport were measured in decibels.
The data collected are shown in the grouped frequency table.

Noise level (d decibels)	$60 \leqslant d < 70$	$70 \leqslant d < 80$	$80 \leqslant d < 90$	$90 \leqslant d < 100$
Frequency	15	16	7	2

a Write down the modal class.

b Use the information in the table to draw a histogram.

c Use your answer to part b to draw a frequency polygon.

3 In a fishing competition the lengths, in centimetres, of all the trout caught were measured. The information collected is shown in the table.

Trout length (*l* cm)	Frequency
$24 \leqslant l < 25$	4
$25 \leqslant l < 26$	14
$26 \leqslant l < 27$	6
$27 \leqslant l < 28$	10
$28 \leqslant l < 29$	6

Draw a frequency polygon for these data.

* **4** The two frequency polygons show the amount of time it took a group of boys and a group of girls to do a crossword puzzle. Who were better at doing the puzzle, boys or girls? Give a reason for your answer.

A03

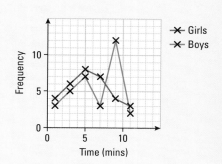

3.7 Drawing and using histograms with unequal class intervals

Objectives

- You can draw a histogram with unequal class intervals.
- You understand frequency density.
- You can find the number of people in a given interval.

Why do this?

If you measure the heights of a number of people, they will cluster around a middle value. Adjusting the size of the class intervals makes these irregularities less noticeable.

Key Points

- In histograms the area of each bar is proportional to the frequency it represents.
- When there are unequal class intervals in a bar you adjust the height by using a scale of **frequency density** rather than width, where

$$\text{frequency density} = \frac{\text{frequency}}{\text{class width}}$$

or frequency = frequency density × class width.

- The area of each bar gives its frequency.

Example 12 The table gives information about the times taken, in seconds, by a number of workers to complete an operation in a factory.

Time taken (t seconds)	Frequency
$10 < t \leqslant 30$	5
$30 < t \leqslant 35$	4
$35 < t \leqslant 40$	8
$40 < t \leqslant 50$	27
$50 < t \leqslant 70$	24

Draw a histogram for these data.

Time taken (t seconds)	Frequency	Class width	Frequency density $= \dfrac{\text{frequency}}{\text{class width}}$
$10 < t \leqslant 30$	5	20	$\frac{5}{20} = 0.25$
$30 < t \leqslant 35$	4	5	$\frac{4}{5} = 0.8$
$35 < t \leqslant 40$	8	5	$\frac{8}{5} = 1.6$
$40 < t \leqslant 50$	27	10	$\frac{27}{10} = 2.7$
$50 < t \leqslant 70$	24	20	$\frac{24}{20} = 1.2$

Work out the width of each class interval (the class width).
Divide the frequency by the class width to find the frequency density which gives the height of each bar.

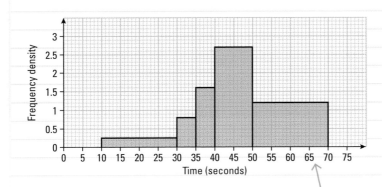

On a grid label the horizontal axis 'Time (seconds)' and the vertical axis 'Frequency density'.
Scale the horizontal axis from 0 to 75 and the vertical axis from 0 to 3.
Draw the bars with no gaps between them.
The first bar goes from 10 to 30 and has a height of 0.25.

Example 13 The histogram gives information about the time, in seconds, taken by students to solve a puzzle.

Time taken (t seconds)	Frequency
$0 < t \leqslant 20$	
$20 < t \leqslant 30$	8
$30 < t \leqslant 40$	
$40 < t \leqslant 50$	
$50 < t \leqslant 70$	

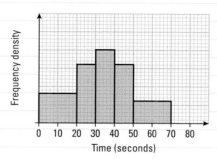

a Complete the frequency table.
b Use the histogram to estimate the number of people who took between 10 and 36 seconds to solve the puzzle.

a Frequency density for $20 < t \leqslant 30$ seconds $= \frac{8}{10} = 0.8$. ← Frequency density $= \dfrac{frequency}{class\ width}$

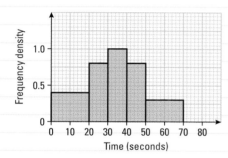

Now put a scale on the histogram.

Frequency = frequency density × class width →

Time taken (t seconds)	Frequency
$0 < t \leqslant 20$	$20 \times 0.4 = 8$
$20 < t \leqslant 30$	8
$30 < t \leqslant 40$	$10 \times 1.0 = 10$
$40 < t \leqslant 50$	$10 \times 0.8 = 8$
$50 < t \leqslant 70$	$20 \times 0.3 = 3$

b

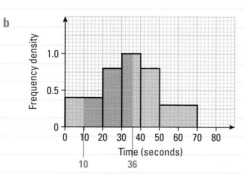

Frequency $= (10 \times 0.4) + (10 \times 0.8) + (6 \times 1.0)$
$= 4 + 8 + 6$
$= 18$ people ← Work out the area between time = 10 and 36 seconds using frequency = frequency density × class width.

Exercise 3H

A **A02**

1 The table gives information about the lifetime of a certain make of torch battery.

Lifetime (*l* hours)	Frequency	Class width	Frequency density
$10 \leqslant l < 15$	4		
$15 \leqslant l < 20$	10		
$20 \leqslant l < 25$	20		
$25 \leqslant l < 30$	15		
$30 \leqslant l < 40$	6		

a Copy and complete the table.

b Draw a histogram for these data.

A02
A03

*2 The table gives information about the distances a group of workers have to travel to work.

Distance (*d* kilometres)	Frequency
$0 < d \leqslant 5$	8
$5 < d \leqslant 10$	16
$10 < d \leqslant 20$	30
$20 < d \leqslant 30$	20
$30 < d \leqslant 40$	6

a Draw a histogram for these data and find an estimate of the number of workers who travel between 15 and 25 minutes.

A☆ **A02**
A03

3 The table gives information about the age of people visiting a theme park one April morning.

a Copy and complete the histogram and table and a scale.

Age (*y* years)	Frequency
$0 < y \leqslant 5$	10
$5 < y \leqslant 10$	28
$10 < y \leqslant 20$	
$20 < y \leqslant 40$	
$40 < y \leqslant 70$	

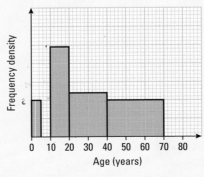

b Find an estimate of how many people between 5 years and 30 years visited the theme park that morning.

3.8 Drawing and using cumulative frequency graphs

Objectives

○ You can construct a cumulative frequency table.
○ You can draw a cumulative frequency graph.

Why do this?

Data are sometimes displayed in a cumulative frequency curve, for example, weights of babies as they get older.

Key Points

● The **cumulative frequency** of a value is the total number of observations that are less than or equal to that value.
● **Cumulative frequency diagrams (graphs)** can be used to find estimates for the number of items up to a certain value.

Example 14 The grouped frequency table shows information about the time, in minutes, taken by 40 runners who had competed in a cross-country race.

Time (t minutes)	Frequency
$t \leqslant 60$	0
$60 < t \leqslant 65$	2
$65 < t \leqslant 70$	12
$70 < t \leqslant 75$	21
$75 < t \leqslant 80$	5

a Draw up a **cumulative frequency table**.
b Draw a cumulative frequency graph.

a

Time (t minutes)	Frequency	Cumulative frequency
$t \leqslant 60$	0	0
$60 < t \leqslant 65$	2	$0 + 2 = 2$
$65 < t \leqslant 70$	12	$2 + 12 = 14$
$70 < t \leqslant 75$	21	$14 + 21 = 35$
$75 < t \leqslant 80$	5	$35 + 5 = 40$

Each time add the frequency to the previous cumulative frequency. The previous frequency was 2 so add the frequency 12 to get the new cumulative frequency 14.

b

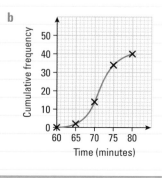

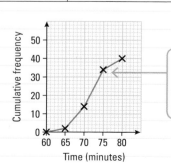

The cumulative frequency 35 for the interval $70 < t \leqslant 75$ is plotted at $(75, 35)$. The plotted points may be joined by a curve or by straight lines.

Example 15 Forty students took a test. The cumulative frequency graph gives information about their marks.

a Use the graph to estimate the number of students who had marks less than or equal to 26.

b Use the graph to work out an estimate for the number of students whose mark was greater than 44.

c 26 students passed the test.
Work out the pass mark for the test.

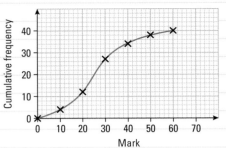

a There are 20 students with a mark less than 26.

b There are 36 students with a mark less than or equal to 44 so there are $40 - 36 = 4$ with a mark greater than 44.

c If 26 pass there will be $40 - 26 = 14$ that fail.
From the graph the pass mark was 22.

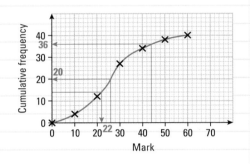

Exercise 3I

B

1 The table shows the ages of people using a bowling alley.

Age (x years)	Frequency	Cumulative frequency
$x \leqslant 10$	3	
$10 < x \leqslant 15$	7	
$15 < x \leqslant 20$	10	
$20 < x \leqslant 25$	15	
$25 < x \leqslant 30$	8	
$30 < x \leqslant 35$	5	
$35 < x \leqslant 40$	2	

a Copy and complete the table.

b Draw a cumulative frequency graph for these data.

2 The cumulative frequency graph shows the time a group of girls spent on school computers.

a Use the cumulative frequency graph to estimate the number of girls who spent up to 4 hours on the computer.

b Use the cumulative frequency graph to estimate the number of girls who spent more than 6 hours on the computer.

c Use the cumulative frequency graph to estimate the number of girls who spent between $3\frac{1}{2}$ and $6\frac{1}{2}$ hours on the computer.

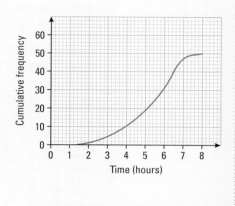

3 The cumulative frequency graph shows the speeds of cars on a motorway.

a Use the cumulative frequency graph to find an estimate for the number of motorists

  i driving at 45 mph or less

  ii driving at between 40 mph and 70 mph.

b How many motorists' speeds were recorded altogether?

c The speed limit on a motorway is 70 mph. Estimate the percentage of cars with a speed greater than 70 mph.

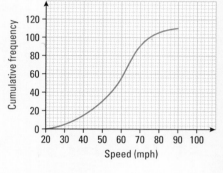

A03

B

3.9 Finding quartiles from a cumulative frequency graph

⊙ Objective

○ You can estimate the median and quartiles from a cumulative frequency graph.

❓ Why do this?

Looking at the age of Britain's population in a frequency table, it is difficult to estimate the median and range. A cumulative frequency graph makes it easy to find the values

✦ Get Ready

Look at this list of numbers: 5, 5, 6, 7, 9, 9, 12, 13, 18, 20, 22, 23

Which numbers are:

a halfway along the list

b three-quarters along the list?

🔍 Key Points

◉ The quartiles divide the frequency into four equal parts.

◉ If there are n values then the quartiles can be estimated from the cumulative frequency graph.

◉ The estimate for the lower quartile is the $\frac{n}{4}$th value.

◉ The estimate for the median is the $\frac{n}{2}$th value.

◉ The estimate for the upper quartile is the $\frac{3n}{4}$th value.

Example 16 The cumulative frequency graph shows information about the times, in minutes, taken by 40 runners who competed in a cross-country race.

    a Find estimates for the median and quartiles.

    b Find estimates for the range and interquartile range.

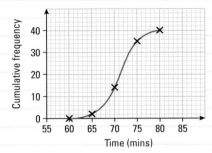

a $Q_1 = 69$ min

     median $= Q_2 = 71.5$ min

           $Q_3 = 73.5$ min

> Q_1 is the $\frac{40}{4} = 10$th value.
> Q_2 is the $\frac{40}{2} = 20$th value.
> Q_3 is the $3 \times \frac{40}{4} = 30$th value.

b   Range $= 80 - 60 = 20$ min

     IQR $= 73.5 - 69 = 4.5$ min

> Range $=$ highest $-$ lowest values
> IQR $= Q_3 - Q_1$

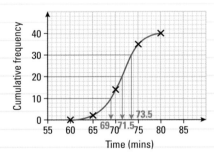

Exercise 3J

1   The cumulative frequency graph shows the scores a group of 100 apprentices got in an engineering examination.

    a Find an estimate for the median (Q_2).

    b Find an estimate for Q_1 and Q_3.

    c Work out the interquartile range.

    d Work out the range.

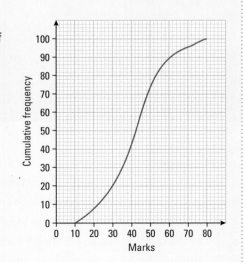

2   The cumulative frequency graph shows the prices of second-hand cars at a garage.

a   Find an estimate for the median (Q_2).

b   Find an estimate for Q_1 and Q_3.

c   Work out the interquartile range.

3   The cumulative frequency graph shows the prices of detached houses on an estate agent's web site.

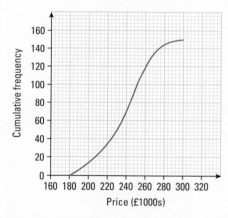

a   Find estimates for the median and quartiles.

b   Find estimates for the range and the interquartile range.

3.10 Drawing and interpreting box plots

Objectives

○ You can construct a box plot given the raw data.
○ You can find the median, quartiles and interquartile range given a box plot.

Why do this?

You can easily show the median and range of data with a box plot. For example, speeds of cars on a section of motorway.

Get Ready

What are the median, lower and upper quartiles, and interquartile range of this list of numbers?

   5   6   6   8   11   13   13   15   17   20   22   25   26   26   29

Key Points

◉ Box plots (sometimes called **box and whisker plots**) are diagrams that show the median, upper and lower quartiles and the maximum and minimum values of a set of data and are often used to compare distributions.

Example 17 | The times run by an athlete had a maximum of 52.1 seconds, a minimum of 47.2 seconds, a median of 48.8 seconds and upper and lower quartiles of 49.3 seconds and 48.2 seconds. Draw a box plot for these data.

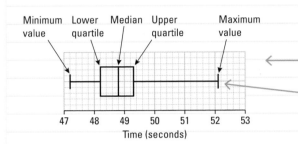

The box shows the spread over the middle 50% of the data (the interquartile range).

The whiskers show the lower 25% and the upper 25% of the data.

Example 18 | The numbers of downloads from a music site during 15 time periods were as follows.

| 5 | 5 | 7 | 12 | 16 | 20 | 21 | 23 |
| 26 | 26 | 27 | 27 | 28 | 29 | 31 |

Draw a box plot for these data.

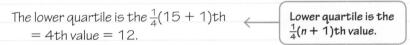

The lowest value is 5 and the highest is 31.

The lower quartile is the $\frac{1}{4}(15 + 1)$th ← *Lower quartile is the $\frac{1}{4}(n + 1)$th value.*
   = 4th value = 12.

The median is the $\frac{1}{2}(15 + 1)$th ← *Median is the $\frac{1}{2}(n + 1)$th value.*
   = 8th value = 23.

Upper quartile is the $\frac{3}{4} \times (15 + 1)$th ← *Upper quartile is the $\frac{3}{4}(n + 1)$th value.*
   = 12th value = 27.

ResultsPlus
Examiner's Tip

These formulae are only for discrete data.

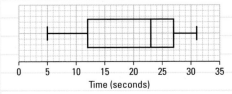

Example 19 The cumulative frequency graphs give information about the number of sales of mobile phones at two shops over 100 days.

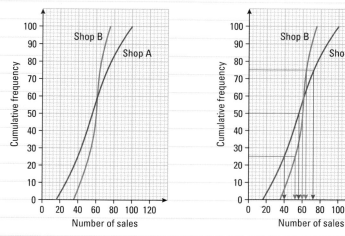

a Draw comparative box plots for these data.
b Compare the sales of the two shops.

Find the maximum and minimum values and the median and quartiles from your graph.

A03

a

	Shop A	Shop B
Least number	16	36
Lower quartile	40	52
Median	56	60
Upper quartile	72	64
Greatest number	100	75

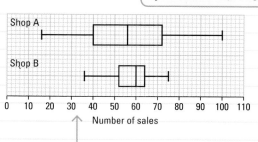

Draw your box plots to the same scale.

b Shop B had a higher median so their sales are generally greater.
Both the range and interquartile range of shop A were greater than those of shop B.
The sales of shop A are more variable from day to day.

Exercise 3K

1 A wildlife park ranger estimated the heights of all the adult giraffes in the park. The tallest was 5.8 metres tall and the shortest was 4.2 metres. The median height was 5 metres, the lower quartile 4.6 metres and the upper quartile 5.6 metres. Draw a box plot for these data.

B

2 The heights of the trees in a small piece of mature woodland were measured in metres. They were as follows.

29   29.2   30.1   32   32.5   34.5   34.5   36.7   38   39.2   39.5   40.0   40.3   40.3   40.4

Draw a box plot for these data.

A

A03 *

3 The cumulative frequency graph gives information about the ages of the male and female members of a cycling club.

   a Use the cumulative frequency diagram to find the quartiles and the maximum and minimum values.

   b Draw two box plots on the same scale using these data and compare and contrast the data.

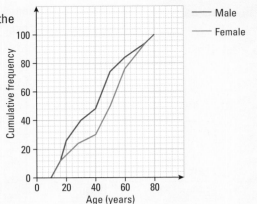

Chapter review

- In a **pie chart** the area of the whole circle represents the total number of items.

- The angles at the centre must add up to 360°.

- The area of each **sector** represents the number of items in that category.
  $$\text{sector angle} = \frac{\text{frequency} \times 360°}{\text{total frequency}} \quad \text{or} \quad \text{frequency} = \frac{\text{sector angle} \times \text{total frequency}}{360°}$$

- A **stem and leaf diagram** is a way of presenting data that makes it easy to see the pattern without losing the actual data.

- A stem and leaf diagram should always have a key.

- From a stem and leaf diagram you can find statistics about the data. The lower quartile (Q_1) is the value a quarter of the way through the data, the second quartile (Q_2) or median is halfway through, and the upper quartile (Q_3) is three-quarters of the way through.

- The **interquartile range** (IQR) is the difference between the upper and lower quartiles = $Q_3 - Q_1$.

- A composite bar chart shows the size of individual categories split into their separate parts.

- A **frequency diagram** for grouped discrete data looks the same as a bar chart except that the label underneath each bar represents a group.

- A **histogram** is similar to a bar chart but because it represents continuous data no gap is left between the bars.

- When drawing a **frequency polygon** you draw a histogram then mark the mid points of the tops of the bars and join these with straight lines.

- More than one frequency polygon can be drawn on the same grid to compare data.

- In histograms the area of each bar is proportional to the frequency it represents.
  $$\textbf{frequency density} = \frac{\text{frequency}}{\text{class width}} \quad \text{or} \quad \text{frequency} = \text{frequency density} \times \text{class width}.$$

- The **cumulative frequency** of a value is the total number of observations that are less than or equal to that value.

- The quartiles divide the frequency into four equal parts and can be estimated from the cumulative frequency graph.

- If there are n values, the estimates are:
  lower quartile = $\frac{n}{4}$th value, median = $\frac{n}{2}$th value, upper quartile = $\frac{3n}{4}$th value.

- Box plots (sometimes called **box and whisker plots**) are diagrams that show the median, upper and lower quartiles and the maximum and minimum values of a set of data and are often used to compare distributions.

Review exercise

1   60 students were asked to choose one of four subjects.
The table gives information about their choices.

Subject	Number of students	Angle
Art	12	72°
French	10	
History	20	
Music	18	

Copy and complete the pie chart to show this information.

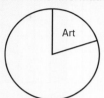

Nov 2008

2   The table gives information about the drinks sold in a café one day.

Drink	Frequency	Size of angle
Hot chocolate	20	80°
Soup	15	
Coffee	25	
Tea	30	

Copy and complete the pie chart to show this information.

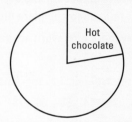

Nov 2008

3   The pie chart gives information about the mathematics exam grades of some students.

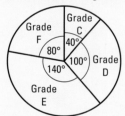

Mathematics exam grades

a   What fraction of the students got grade D?

Eight of the students got grade C.

b   i   How many of the students got grade F?   ii   How many students took the exam?

This accurate pie chart gives information about the English exam grades for a different set of students.

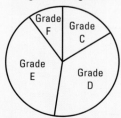

English exam grades

Sean says 'More students got a grade D in English than in mathematics.'

c   Sean could be **wrong**. Explain why.

June 2008

4 Mr White recorded the number of students absent one week.
The dual bar chart shows this information for the first four days.

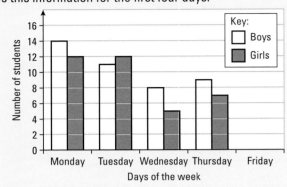

a How many boys were absent on Monday?
b How many girls were absent on Wednesday?

On Friday, 9 boys were absent and 6 girls were absent.
c Use this information to complete the bar chart.

On only one day more girls were absent than boys.
d Which day?   *March 2008*

5 Zoe recorded the weights, in kilograms, of 15 people. Here are her results.

    87 51 46 77 74 58 68 78 48 63 52 64 79 60 66

a Draw a diagram to show these results.
b Write down the number of people with a weight of more than 70 kg.
c Work out the range of the weights.   *March 2009 (amended)*

6 Jason collected some information about the heights of 19 plants.
This information is shown in the stem and leaf diagram.

    1 | 1   2   3   3
    2 | 3   3   5   9   9
    3 | 0   2   2   6   6   7
    4 | 1   1   4   8   Key 4|8 means 48 mm

Find the median.   *Nov 2008*

7 The table shows some information about the weights (w grams) of 60 apples.
On a copy of the grid, draw a frequency polygon
to show this information.

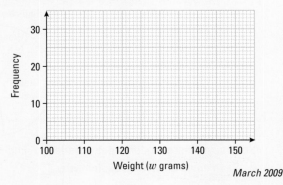

Weight (w grams)	Frequency
$100 \leqslant w < 110$	5
$110 \leqslant w < 120$	9
$120 \leqslant w < 130$	14
$130 \leqslant w < 140$	24
$140 \leqslant w < 150$	8

March 2009

8 60 students take a science test.
The test is marked out of 50.
This table shows information about the students' marks.

Science mark	0–10	11–20	21–30	31–40	41–50
Frequency	4	13	17	19	7

On a copy of the grid, draw a frequency
polygon to show this information.

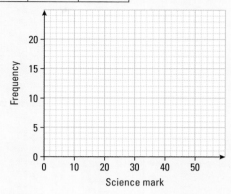

June 2008

* 9

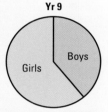

Pie chart showing proportion
of boys and girls in Year 9

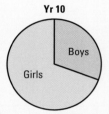

Pie chart showing proportion
of boys and girls in Year 10

To draw the pie chart for boys and girls in Years 9 and
10 combined, Kimberly drew the pie chart on the right:

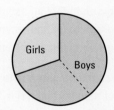

James said that this could not be correct.
Explain who is right.

Pie chart showing proportion of boys
and girls in Year 9 and Year 10

* 10 John and Peter each own a garage. They both sell used cars.
The box plots show some information about the prices of cars at their garages.

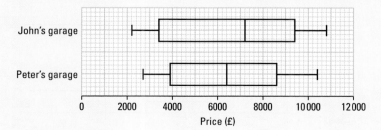

Compare the distribution of the prices of cars in these two garages.
Give **two** comparisons.

Nov 2008

B

11 The cumulative frequency graph shows some information about the ages of 100 people.

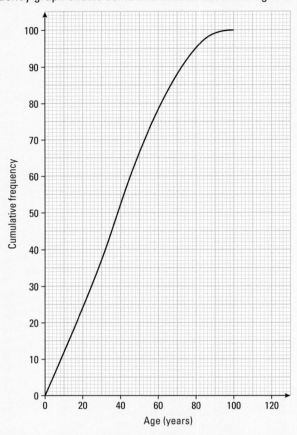

a Use the graph to find an estimate for the number of these people less than 70 years of age.

b Use the graph to find an estimate for the median age.

c Use the graph to find an estimate for the interquartile range of the ages.

Nov 2008

A03

12 Verity records the heights of the girls in her class.

The height of the shortest girl is 1.38 m.

The height of the tallest girl is 1.81 m.

The median height is 1.63 m.

The lower quartile is 1.54 m.

The interquartile range is 0.14 m.

a Using this scale, draw a box plot for this information.

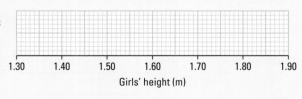

The box plot shows information about the heights of the boys in Verity's class.

b Compare the distributions of the boys' heights and the girls' heights.

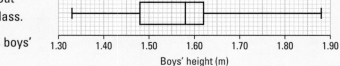

March 2008

*13 Lucy did a survey about the amounts of money spent by 120 men during their summer holidays. The cumulative frequency table gives some information about the amounts of money spent by the 120 men.

A survey of the amounts of money spent by 200 women during their summer holidays gave a median of £205. Compare the amounts of money spent by the women with the amounts of money spent by the men.

Amount (£A) spent	Cumulative frequency
$0 < A \leq 100$	13
$0 < A \leq 150$	25
$0 < A \leq 200$	42
$0 < A \leq 250$	64
$0 < A \leq 300$	93
$0 < A \leq 350$	110
$0 < A \leq 400$	120

May 2009

14 The box plot gives information about the distribution of the weights of bags on a plane.

a Jean says the heaviest bag weighs 23 kg.
   She is **wrong**. Explain why.

b Write down the median weight.

c Work out the interquartile range of the weights.
There are 240 bags on the plane.

d Work out the number of bags with a weight of 10 kg or less.

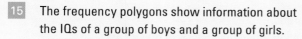

June 2009

15 The frequency polygons show information about the IQs of a group of boys and a group of girls.

a Write down an estimate for the number of girls with an IQ of 110.

b Write down an estimate for the number of boys with an IQ of 110.

c Use the frequency polygon to compare the overall IQs of the boys and the girls.

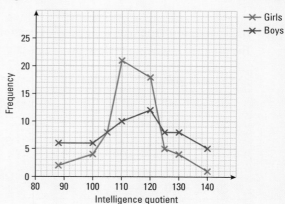

16 The table gives some information about the lengths of time some boys took to run a race. Draw a histogram for the information in the table.

Time (t minutes)	Frequency
$40 \leq t < 50$	16
$50 \leq t < 55$	18
$55 \leq t < 65$	32
$65 \leq t < 80$	30
$80 \leq t < 100$	24

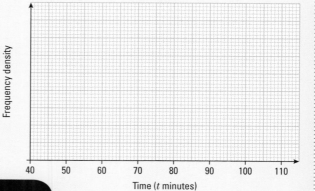

ResultsPlus
Exam Question Report

73% of students answered this sort of question poorly.

March 2009

A

17 On Friday, Peter went to the airport.
He recorded the number of minutes that each plane was delayed.
He used his results to work out the information in this table.

	Minutes
Shortest delay	0
Lower quartile	2
Median	8
Upper quartile	18
Longest delay	41

a Using this scale, draw a box plot to show the information in the table.

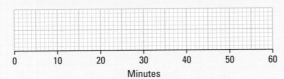

Peter also went to the airport on Saturday.
He recorded the number of minutes that each plane was delayed.
The box plot below was drawn using this information.

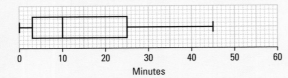

A03

b Comment on the plane delays.

March 2009 (amended)

A02

18 The speeds of 100 cars on a motorway were recorded.
The grouped frequency table shows some information about the speeds of these cars.

Speed (s mph)	Frequency
$40 < s \leqslant 50$	4
$50 < s \leqslant 60$	19
$60 < s \leqslant 70$	34
$70 < s \leqslant 80$	27
$80 < s \leqslant 90$	14
$90 < s \leqslant 100$	2

a On a copy of the grid, draw an appropriate graph for your table.

b Find an estimate for the median speed.

c Find an estimate for the interquartile range.

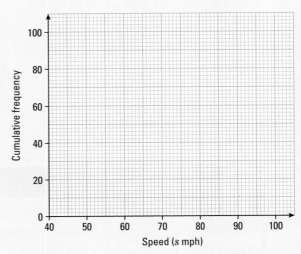

June 2008 (amended)

19 The incomplete histogram and table give some information about the distances some teachers travel to school.

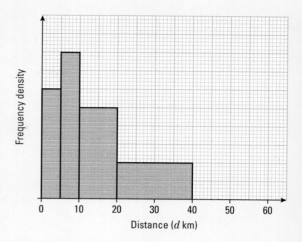

a Use the information in the histogram to complete the frequency table.

Distance (d km)	Frequency
$0 < d \leqslant 5$	15
$5 < d \leqslant 10$	20
$10 < d \leqslant 20$	
$20 < d \leqslant 40$	
$40 < d \leqslant 60$	10

b Use the information in the table to complete the histogram.

Nov 2008

20 The table gives information about parcel sizes and their frequency.

Weight (w kg)	Frequency	Frequency density
$0 < w \leqslant 5$	20	
$5 < w \leqslant 15$	30	
$15 < w \leqslant 25$	15	
$25 < w \leqslant 35$	10	
$35 < w \leqslant 40$	5	

a Copy and complete the table.

b Draw a histogram for these data.

The weight limit for parcels going by Royal Mail is 20 kg.

c Work out an estimate for the number of parcels which will weigh 20 kg or less.

d Work out an estimate for the number of parcels weighing between 10 and 30 kg.

21 The histogram shows information about the lifetime of some batteries.

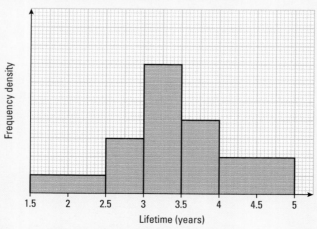

Two of the batteries had a lifetime of between 1.5 and 2.5 years.
Find the total number of batteries.

June 2008

4 LINE DIAGRAMS AND SCATTER GRAPHS

Life expectancy over time is one variable often represented using a line graph. The line for life expectancy in the UK shows a continual increase from 1980 to the present day. In 1980, a man could expect to live to an age of about 71 years whilst the average life expectancy for a woman was 77. By 2009, the life expectancy for both sexes had gone up considerably with average life expectancy for a baby girl at 81.5 years and for a baby boy at 77.2 years.

◎ Objectives

In this chapter you will be able to do the following for various data types:
- produce and interpret line graphs
- produce and interpret scatter graphs
- see if there is any linear association between two variables
- distinguish between positive, negative and zero correlation
- draw lines of best fit
- use a line of best fit to predict values of a variable.

◈ Before you start

You need to:
- understand how to draw, label and scale axes
- substitute numbers in simple algebraic expressions.

4.1 Plotting points on a graph

4.1 Plotting points on a graph

◉ Objective

● You can plot and interpret graphs that model real situations.

❓ Why do this?

There are a number of situations in which varying one thing causes another to vary. For example, as you increase the amount of electricity you consume the price you have to pay for it increases.

Plotting points on a graph

Key Points

● The **coordinates** of a point P are written (x, y) where x is the distance across the graph and y is the distance up the graph.

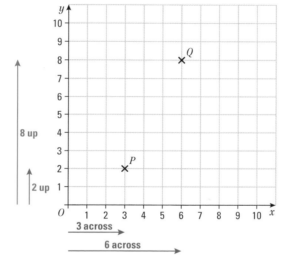

Point P is $(3, 2)$. The x-coordinate comes first and the y-coordinate comes second.

Point Q is $(6, 8)$.

Plotting the graph of a formula

Key Points

● To plot the graph of a formula, find pairs of corresponding values and use them as coordinates for the graph.

Example 1

a Plot the graph of $y = 2x + 1$ for values of x between 0 and 5.

b Find the value of y when $x = 2.5$ using your graph.

c Find the value of x which gives a value of 10 to y.

a

x	0	1	2	3	4	5
y	1	3	5	7	9	11

← You let x take the values 0, 1, 2, 3, 4, 5 and work out the corresponding values of y.

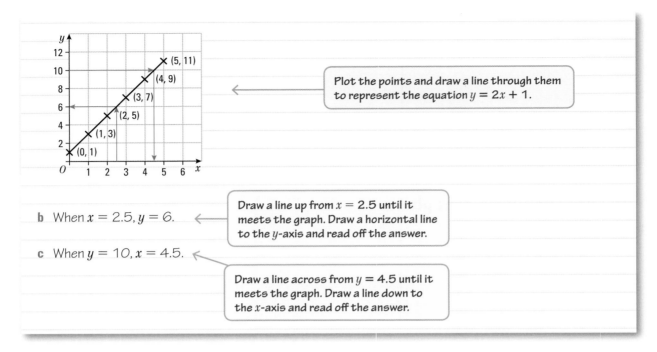

Plot the points and draw a line through them to represent the equation $y = 2x + 1$.

b When $x = 2.5$, $y = 6$.

Draw a line up from $x = 2.5$ until it meets the graph. Draw a horizontal line to the y-axis and read off the answer.

c When $y = 10$, $x = 4.5$.

Draw a line across from $y = 4.5$ until it meets the graph. Draw a line down to the x-axis and read off the answer.

Real-life graphs

Example 2

This graph shows the number of pounds you get for a number of US dollars ($) at an exchange rate of $1 = £0.60.
Use the graph to find out
a how many pounds you would get for $7
b how many dollars you would get for £3.

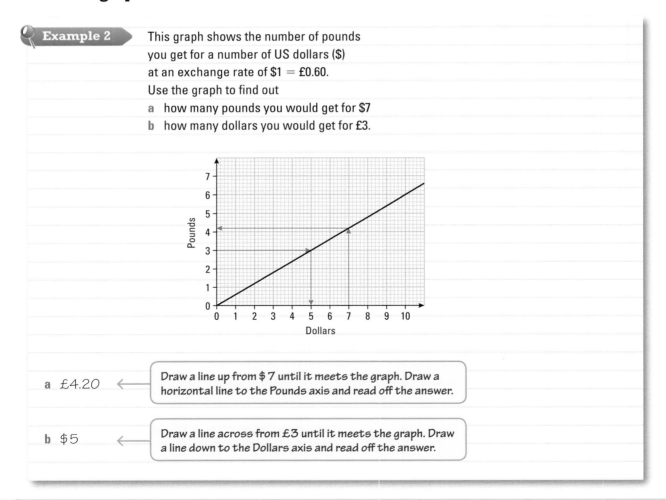

a £4.20

Draw a line up from $7 until it meets the graph. Draw a horizontal line to the Pounds axis and read off the answer.

b $5

Draw a line across from £3 until it meets the graph. Draw a line down to the Dollars axis and read off the answer.

Example 3

Use the graph to find the metric equivalent of 3.8 miles and the distance in miles that corresponds to a distance of 11.2 km.

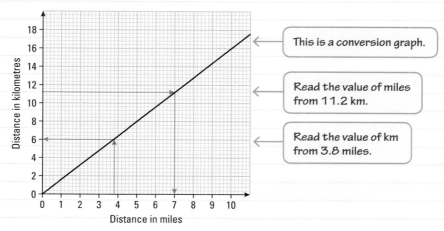

This is a conversion graph.

Read the value of miles from 11.2 km.

Read the value of km from 3.8 miles.

3.8 miles = 6 km and 11.2 km = 7 miles.

Example 4

This graph shows the relationship between length of side and area for squares.

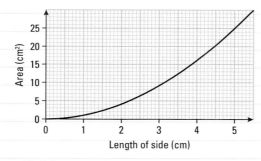

a Find the area of a square whose sides are 3.6 cm long.

b Find the length of sides of a square that has an area of 2 cm².

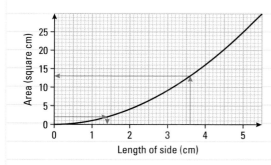

a 13 cm²

b 1.4 cm

✷ Exercise 4A   Questions in this chapter are targeted at the grades indicated.

1   This graph shows the number of pounds you get
    for a number of euros.

    a Estimate the number of pounds you can get for
      7.75 euros.

    b Estimate the number of euros you can get for £5.00.

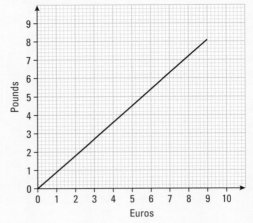

2   This graph shows the relationship between the diameter of
    an orange and its surface area.

    a Estimate the surface area if the diameter is 5.5 cm.

    b Estimate the diameter if the surface area is 50 cm².

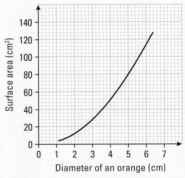

3   a Copy and complete the following table.

    b Plot the graph of the line $y = 2x + 1$
      for values of x between 0 and 5.
      Use values 0 to 6 on the x axis and 0 to 12 on the y axis.

    c Use your graph to find the value of y when x is 2.5.

    d Use your graph to find the value of x when y is 8.

x	0	1	2	3	4	5
$y = 2x + 1$	1	3	5			

4.2 Straight-line graphs

◎ Objectives

● You can recognise and plot equations
  that correspond to straight lines
  graphs, including gradients.

❓ Why do this?

If you do an experiment to see how far a spring stretches when
different weights are put on it, you should be able to predict
from your experiment how much it stretches for a given weight,
and what weight would make it stretch a certain amount.

⬆ Get Ready

1. In the formula $y = x + 3$, find y for the following values of x.
   a $x = 0$ b $x = 2$ c $x = 5$
2. Plot the graph of $y = x + 3$ for values of x between 0 and 5.

Key Points

◉ The general form for the **equation of a line** graph is $y = mx + c$.

◉ The constant m is the **gradient** of a straight line graph. This is the amount the y value increases for an increase of 1 in the x value.

◉ The value of the gradient may be found using gradient $= \dfrac{\text{increase in } y}{\text{increase in } x}$

◉ When $x = 0$, $y = c$. The value of c is the value at which the line cuts the y-axis. This is known as the y-intercept.

Example 5 Draw a graph of the equation $y = 1.5x + 2$ for values from $x = 0$ to $x = 4$

When $x = 0$, $y = 2$
When $x = 4$, $y = 1.5 \times 4 + 2 = 8$ ← These two points fix the position of the line.

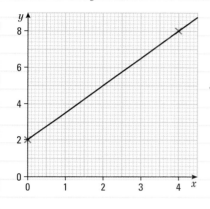

← The line is drawn between them.

Example 6 The graph shows a straight-line graph, representing the equation $y = mx + c$, where m and c are fixed numbers.
Find the equation of this line.

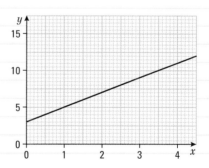

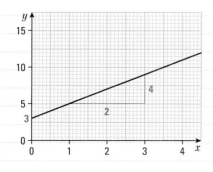

$c = 3$ ← When $x = 0$, $y = 3$

$m = \dfrac{4}{2} = 2$ ← When x increases by 2, y increases by 4.
$\quad\quad\quad\quad\quad m = \dfrac{\text{increase in } y}{\text{increase in } x}$

$y = 2x + 3$

Exercise 4B

D

1 Find the gradient of the line.

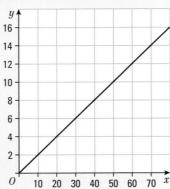

2

x	2	3	4	5	6	7
y	4	10	16	22	28	34

a Draw axes for 0 to 8 on the x axis and 0 to 40 on the y axis.

b Plot the points shown in the table on these axes.

c Join the points with a straight line.

d Work out the gradient of the line.

A03

3 The graph represents the relationship between the tail length (cm) and the body length (cm) of a small mammal.

x	0	1	2	3	4	5
$y = 2x + 1$	1	3	5			

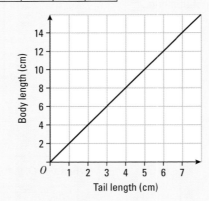

Write down the gradient of the line and interpret it in context.

C

A03

* 4 The gradient of a line that represents the mass in grams of piglets y, and the number of piglets in a litter, x, is -50.

What does this mean in terms of litter size and piglet mass?

4.3 Drawing and using line graphs

Objectives

○ You can draw line graphs.
○ You can estimate values from a line graph.
○ You can plot and interpret real-life graphs.

Why do this?

Line graphs enable you to see how one variable changes as another related one changes. For example, how the temperature changes as the time of day changes.

Get Ready

Look at this set of axes.
What does one small division represent on
a the x-axis
b the y-axis?

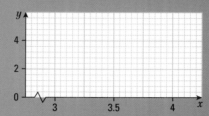

Key Points

◉ Sometimes when sampling you take two observations from each selected member of the population. We call these bivariate data.
◉ Bivariate data consists of pairs of related variables.
◉ Pairs of observations can be plotted on a **line graph**.
◉ Time is often the variable along the horizontal axis in line graphs.

Example 7

The table gives information about the close-of-day price for buying a share in a particular company during one week in June.

Day (June)	Wed 13th	Thur 14th	Fri 15th	Sat 16th	Sun 17th	Mon 18th	Tues 19th
Price per share (pence)	138	144	141	Closed		137	143

    a Draw a line graph for these data.
    b On which day was the share price at its highest?
    c On which day was the share price at its lowest?

a

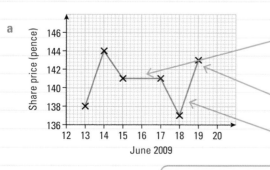

You often get periods like this when nothing happens. Monday's price starts at closing price of Friday.

Plot the points on the graph.

Join the points with straight lines.

b 14th March ← Find when the highest value occurs.

c 18th March ← Find when the lowest value occurs.

Example 8 The line graph shows the temperature, in °C, at different times of the day during a day in April.

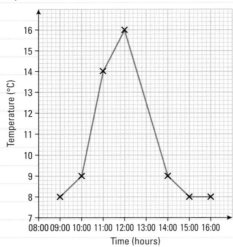

The temperature for 13:00 hours was missed.

a Estimate the temperature at 13:00 hours.

b Estimate the times when the temperature was 8.5°C.

c What was the highest temperature reached and at what time did it occur?

d In which hour did the temperature rise most quickly?

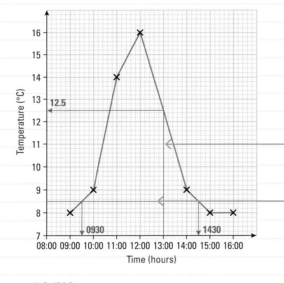

To estimate the temperature at 13:00 hours, draw a line up from 13:00 until it meets the line graph. Draw a horizontal line from here to meet the vertical axis. Read off the answer.

To estimate the times at which the temperature was 8.5°C draw a line horizontally from 8.5. Draw lines down from where this crosses the line graph. Read off the answers.

a 12.5°C

b 09:30 and 14:30 hours. ← It is 8.5°C twice during the day.

c 16°C at 12:00 hours. ← This will be the highest point.

d The temperature rose most quickly between 10:00 and 11:00 hours. ← The steeper the gradient of the line, the more quickly the temperature changed.

Exercise 4C

1   The line graph shows the time it took a parent to take her child to school.

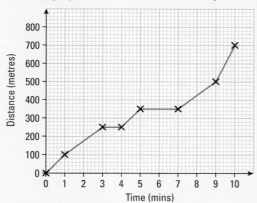

a   How far was the journey to school?

b   How many times did they stop on the journey and for how long did they stop?

c   Use the line graph to estimate how long it took them to walk 600 metres.

d   Use the line graph to estimate how far they had walked in 8 minutes.

2   The line graph shows the depth, in metres, of water in a reservoir each month for one year.

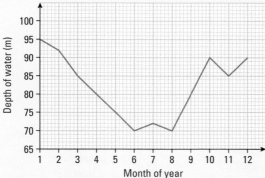

a   In which month was the water at its deepest? What was the depth?

b   Use the line graph to estimate the depth of water in the fourth month.

c   In which months was the water at its lowest? Suggest a reason for this.

A03

3   The line graph shows the power supplied, for domestic power consumption, by Yorkoft power station over a 12-hour period in September.

A03

a   At what time was domestic consumption of power at its highest?
    Suggest a reason why demand is high at this time.

b   What was the consumption at 16:00 hours?

c   Use the line graph to find an estimate of the times when the consumption was 150 000 kilowatts.

d   Write down the time when consumption was at its lowest. Suggest a reason for this.

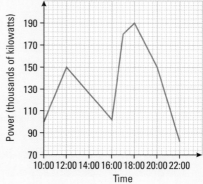

4.4 **Drawing and using scatter graphs**

⊙ **Objectives**

- You can use a scatter graph to see if there is any relationship between pairs of variables.
- You can plot and interpret real-life graphs.

⊙ **Why do this?**

You could use a scatter graph to plot the data on local speed limits and the number of children involved in traffic accidents, and see if there is a relationship between the two.

⊕ **Get Ready**

What are the values of points A to E?

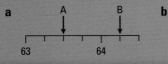
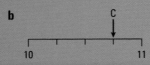
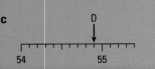
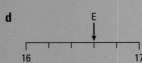

🔍 **Key Points**

- When taking pairs of observations from members of a sample, we plot points on a graph to see if there is any relationship between the two variables being observed.
  The resulting graph is called a scatter diagram or **scatter graph**.
- A scatter graph enables you to see how scattered pairs of values are when plotted.

🔍 **Example 9**

In a certain city council area the proportion of open spaces was 2% and the percentage of accidents involving children was 40%.
The table shows the figures for seven other council areas.

Open space (%)	5	1.4	2.5	5.2	12.2	15	6.3
% accidents involving children	43	40	36	33	30	25	32

a Draw a scatter diagram for these data.
b Describe how the variables are related.

a

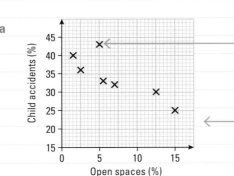

> The pairs of values are plotted on the graph in the usual way. This cross shows open spaces 5%, child accidents 43%.

> There are two variables, open spaces and child accidents.

b We can see that the crosses are in a roughly downward sloping line.
So there seems to be a relationship between the amount of open spaces and the percentage of accidents involving children.
The relationship is not perfect but a general trend can be seen.
The greater the percentage of open space, the fewer the children involved in accidents.

Example 10   The table below shows females' years of birth and the age they could expect to live to.

Year of birth	1986	1991	1996	2001	2006
Life expectation (years)	77.7	78.7	79.4	80.4	81.5

Data source: Stats.gov.uk

   a Draw a scatter graph for these data.

   b Comment on the relationship between birth year and life expectancy.

a

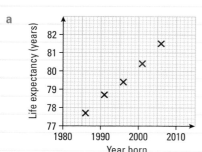

b There seems to be a close relationship.
    The later a person was born the greater their life expectancy at birth.

4.5 Recognising correlation

Objectives

- You can distinguish between positive, negative and zero correlation.
- You appreciate that correlation is a measure of the strength of the association.

Why do this?

If a scatter diagram shows points that seem to have a relationship then they are said to be correlated. The year you were born and your life expectancy are correlated.

Key Points

- If every time one variable changes the other variable changes as well, we say the variables are correlated. If the points lie almost in a straight line they are said to be linearly correlated. Linear means 'in a straight line'. A relationship between pairs of variables is called a **correlation**. We will only consider linearly correlated variables here.

- If one variable increases as the other one increases the correlation is said to be positive.

- If one variable decreases as the other increases the correlation is said to be negative.

- If there is no relationship between the variables then there is no correlation and the correlation is said to be zero.

These three possibilities are shown in these graphs.

Positive correlation

As one value increases the
other one increases.

Negative correlation

As one value increases the
other decreases.

No correlation

The points are random and
widely spaced.

- If there is perfect **positive correlation** between two variables the correlation is given a value of $+1$.
- If there is perfect **negative correlation** between two variables the correlation is given a value of -1.
- If there is **no correlation** then the correlation is zero and is given the value 0.

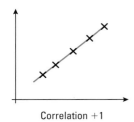

Correlation $+1$

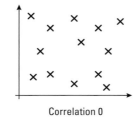

Correlation 0

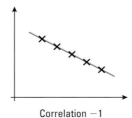

Correlation -1

Example 11

A bar is supported at each end. A weight is hung in the middle and the amount that the middle of the bar sags is measured.
The scatter diagram shows the resulting sag for different weights.

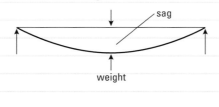

a Describe the correlation.
b Describe the relationship between the load and the amount of sag in the middle of the bar.

a Positive correlation. ← As one variable increases so does the other one.

b The greater the weight the greater the sag.

Example 12

Which of the following pairs of variables are related?
a A child's height and weight
b A car's engine size and the number of seats
c The number of MP3 players and the number of flat screen televisions sold by a store
d The number of speed cameras and the number of speeding fines

The variables of **a** and **d** are related.

The number of seats in a car doesn't affect the size of the engine. Sales of MP3 players don't affect sales of flat screen televisions.

Exercise 4D

1 The table gives information about the engine size, in litres, and the petrol consumption, in miles per litre, of eight cars.

Engine size (litres)	1.6	2.6	1.4	1.0	2.2	1.2	3.0	1.7
Petrol consumption (mpl)	10	7	11	12	8	13	6	9

a Draw a scatter graph for these data.

b Describe the correlation.

c Describe the relationship between engine size and petrol efficiency.

2 The table gives information about the selling price of computers and their screen size.

Screen size (inches)	10	17	17	15	12	14	16	11	16
Selling price (£)	700	700	550	400	450	480	300	350	880

a Draw a scatter graph for these data.

b Describe the correlation.

c What can you say about the relationship between screen size and price?

3 In the year 2000 a river was restocked with fish.
The estate with the fishing rights kept a record of the number of fish caught on a particular stretch of the river for the next 10 years. The data collected are shown in the table.

Year after restocking	1	2	3	4	5	6	7	8	9	10
Number of fish caught	180	165	168	155	158	150	145	148	140	135

a Draw a scatter graph for these data.

b Describe the correlation.

c Describe the relationship between years after restocking and number of fish caught.

4 Which of the following pairs of variables are related? Give a reason for your answer.

a A car's maximum speed and its weight

b The length of a motorway and the number of petrol stations on the motorway

c The number of washing machines and the number of computers sold by an electrical store

d The number of bicycles and the number of cycle helmets sold by a shop

4.6 Drawing lines of best fit

- You can draw lines of best fit by eye.
- You can explain an isolated point on a scatter diagram.

❓ Why do this?

A diver could plot a graph of pressure at certain depths, with a line of best fit and use the line to estimate a formula relating depth to pressure.

🔍 Key Points

- If the points on a scatter graph lie approximately in a straight line the correlation is said to be linear.

- If the points are roughly in a straight line you can draw a **line of best fit** through them.

- A line of best fit is a straight line that passes as near as possible to the various points so as to best represent the **trend** of the graph.

- A line of best fit does not have to pass through any of the points, but it may pass through some of them.

- When drawing a line of best fit, draw it so that roughly the same numbers of points are either side of the line and so that the line drawn best represents the trend of the points.

- If lines of best fit are added to the scatter graphs in Example 9 and Example 10 they will look like this.

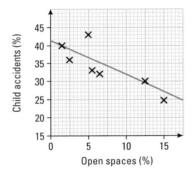

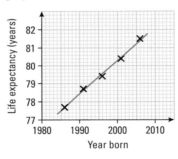

- An isolated point is an extreme point that lies outside the normal range of values.

- When drawing lines of best fit, or reaching conclusions, isolated points should be omitted from your data set.

🔶 A03

🏅 Example 13 The table shows information about the percentage of people unemployed in a country and the percentage rise in wages over a number of years.

Unemployed (%)	2.0	2.2	3.0	1.8	1.6	4.0	1.6	1.8	2.0	1.4
Rise in wages (%)	2.7	3.0	2.7	3.5	1.4	1.6	4.2	3.5	3.8	3.9

a Draw a scatter graph of these data.

b Describe the correlation between the percentage of unemployed and the rise in wages.

c Draw a line of best fit on your scatter graph.

An economist thinks that when there is a lot of unemployment wage rises will be lower.

d Is the economist right? Give reasons.

a

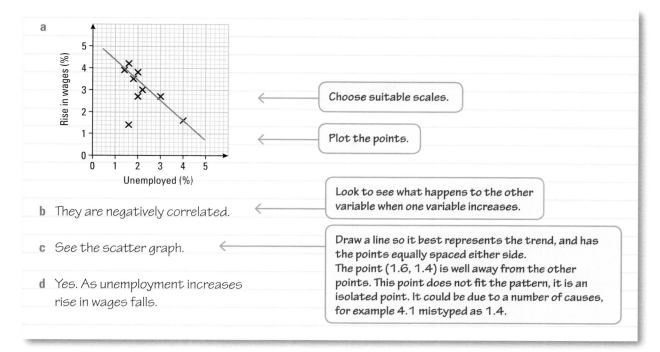

Choose suitable scales.

Plot the points.

b They are negatively correlated.

Look to see what happens to the other variable when one variable increases.

c See the scatter graph.

d Yes. As unemployment increases rise in wages falls.

Draw a line so it best represents the trend, and has the points equally spaced either side.
The point (1.6, 1.4) is well away from the other points. This point does not fit the pattern, it is an isolated point. It could be due to a number of causes, for example 4.1 mistyped as 1.4.

Exercise 4E

1 The scatter diagram shows the energy consumption and the GNP (Gross National Product, a measure of economic prosperity) for nine countries.

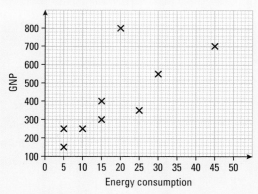

a Copy the scatter diagram.

b One point seems to be an isolated point. Circle the isolated point and write down its coordinates.

c Ignoring the isolated point, draw a line of best fit on your diagram.

2 The table shows the ages and prices of 10 second-hand cars.

Age (years)	1	2	3	3	5	2	7	8	9	12
Price (£1000)	10	7.5	7	6.5	4.5	8	2	1.5	1	0.5

a Draw a scatter graph for these data.

b One point seems to be an isolated point. Circle it and suggest a reason why this might have occurred.

c Ignoring the isolated point, draw a line of best fit on your scatter graph.

D

3 The table shows information that has been recorded by a researcher about the mean high and the mean low temperatures for some cities. The first six cities have been plotted on the scatter graph.

	Amsterdam	Berlin	Mumbai	Dublin	Hong Kong	London
High temp.	54	55	87	56	77	58
Low temp.	46	40	74	42	68	44

	Madrid	Oslo	Ottawa	Paris	Rangoon	Rome
High temp.	65	50	32	59	89	71
Low temp.	54	35	51	43	73	51

a Copy the scatter diagram and complete it by plotting the last six points.

b There is one point that seems to be an isolated point. Circle this point.

c Write down the name of the city that is the isolated point. Suggest a reason for this isolated point.

d Ignoring the isolated point, draw a line of best fit on your scatter graph.

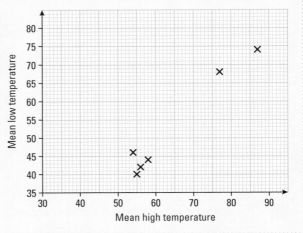

4.7 Using lines of best fit to make predictions

◎ Objectives

● You can use a line of best fit to predict a value of one of a pair of variables given a value for the other variable.

⊘ Why do this?

A graph with a line of best fit for the number of drinks that a shop sold in the spring, at cooler temperatures, might help the shop predict how many drinks they might sell in the summer, as temperatures increase.

🔵 Key Points

● If a value of one of the variables is known, you can estimate the corresponding value of the other variable by using the line of best fit.

● For example, to estimate the likely rise in wages given that 3.6% were unemployed, you draw a vertical line at 3.6% until it hits the line of best fit. You then draw a horizontal line from there and read off where it comes on the vertical scale. In this case, you read off 2%.

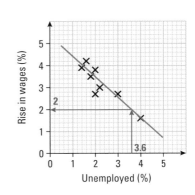

● Using lines of best fit when the value you are finding is within the range of values on the scatter diagram is called **interpolation** which is usually reasonably accurate.

● Using the line of best fit to find values outside the range of values on the scatter diagram is called **extrapolation** which may not be very accurate.

Example 14 The scatter graph gives information about the population density, in people per hectare, and the distance, in kilometres, from a city centre.

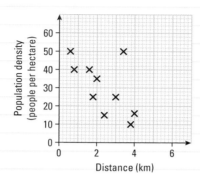

a Identify any possible outliers.

b Draw a line of best fit.

c Estimate the population density at 1.2 km from the centre.

d Estimate how far an area with a density of 30 people per hectare is from the centre.

a Point (3.4, 50) is an isolated point.

b

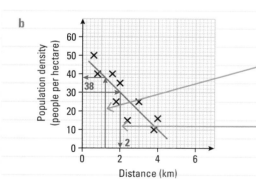

For **c** draw a line from 1.2 on the horizontal axis up to the line of best fit. From where it hits the line draw a horizontal line across to the vertical axis and read off the required value.

For **d** draw a horizontal line from 30 on the vertical axis across to the line of best fit. From where it hits the line draw a vertical line down to the horizontal axis and read off the required value.

c 38 people per hectare.

d 2 km

ResultsPlus

Examiner's Tip

Always draw the lines on your diagram. Even if you get the wrong answer you might get marks for the correct method.

Exercise 4F

1 The scatter diagram shows the marks in statistics and mathematics of a group of students.

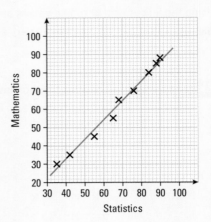

a A student gets a mark of 50 in statistics.
Use the line of best fit to find the mark he is likely to get in mathematics.

b A student gets a mark of 60 in mathematics.
Use the line of best fit to find the mark he is likely to get in statistics.

2 The scatter graph shows the latitudes and the mean highest temperatures of 10 countries of the world.

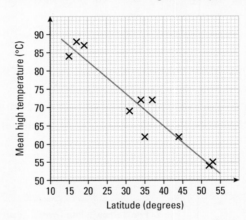

a Use the line of best fit to work out an estimate of the mean high temperature at a latitude of:
  i 40 degrees   ii 25 degrees.

b Use the line of best fit to work out an estimate of the latitude where you are likely to get a mean high temperature of:
  i 60°C   ii 85°C.

3 An engineer measured the length of a copper rod at various temperatures.
The table represents the data collected.

Temperature (°C)	20.5	27.5	40	43	55	60	65	70
Length in metres	2.4611	2.4614	2.4619	2.462	2.4623	2.4629	2.463	2.4636

a Draw a scatter graph for this data.

b Work out an estimate of the temperature when the length is
  i 2.462 metres
  ii 2.463 metres.

c Work out an estimate of the length when the temperature is 30°C.

Chapter review

- Bivariate data consists of pairs of related variables.
- Pairs of observations can be plotted on a line graph.
- A **scatter graph** enables you to see how scattered pairs of points are when plotted.
- A relationship between pairs of variables is called a **correlation**.
- If one variable increases as the other one increases, the correlation is said to be positive.
- If one variable decreases as the other increases, the correlation is said to be negative.
- If there is no relationship between the variables then there is **no correlation** and the correlation is said to be zero.
- A **line of best fit** is a straight line that passes as near as possible to the various points so as to best represent the **trend** of the graph.
- An isolated point is an extreme point that lies outside the normal range of values.
- If a value of one of the variables is known, you can estimate the corresponding value of the other variable by using the line of best fit.

Review exercise

1 The populations, in millions, of a capital city between 1951 and 2001 are shown on the line graph.

   a Estimate the population in 1995.

   b Estimate the year when the population was 7.6 million.

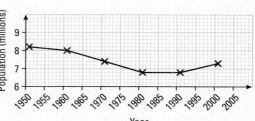

A03

2 The table gives information about the number of people who were unemployed in a seaside town over the course of a year.

A03 **D**

Month	Jan	Feb	Mar	Apr	May	Jun	Jul	Aug	Sep	Oct	Nov	Dec
No. of unemployed	110	98	56	50	48	34	40	30	45	–	85	105

   a Plot a line graph for these data.

   b Estimate the number of unemployed people in October.

   c For which month of the year was the number of unemployed lowest? Give a reason for this.

3 The table gives some information about the science and art marks of some students.

Student	A	B	C	D	E	F	G	H	I	J
Science mark	78	65	48	68	89	95	46	38	56	70
Art mark	45	58	60	50	43	70	52	58	50	50

   a Draw a scatter graph for these data.

   b One point appears to be an isolated point. Circle that point.

   c Ignoring the isolated point, describe the correlation.

A03

   d Describe the relationship between the science marks and the art marks.

D

4 The table shows the height, in metres, above sea level and the temperature, in °C, at 06:00 hours at 10 places in Austria on one day in July.

Height (100s metres)	11	15	10	5	4	5	9	12	18	16
Temperature (°C)	11	6	10	16	14	13	8	9	5	6

a Draw a scatter graph for these data.

b Describe the correlation.

A03 c Describe the relationship between the height and the temperature.

d Draw a line of best fit on your scatter graph.

A02
A03

5 The table gives some information about the length of a metal rod at different temperatures.

Temperature (x °C)	60	65	70	75	80	85
Length (y mm)	90.2	90.8	91.7	92	93	94.2

a Draw a scatter diagram for these data. Use values from 90 to 94 for the y axis and 55 to 85 for the x axis.

b Describe the correlation.

c Draw a line of best fit on your graph.

d Work out an estimate for the gradient of the line of best fit.

e Interpret the gradient in terms of length and temperature.

A03 **6** A superstore sells the Clicapic digital camera.

The price of the camera changes each week.

Each week the manager records the price of the camera and the number of cameras sold that week.

The scatter graph shows this information.

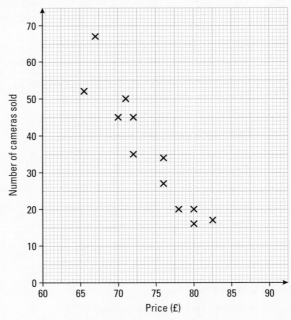

a Describe how the price of the camera and the number of cameras sold are related.

b Draw a line of best fit on a copy of the scatter graph.

Nov 2008

7 The scatter graph shows some information about the ages and values of fourteen cars.
The cars are the same make and type.

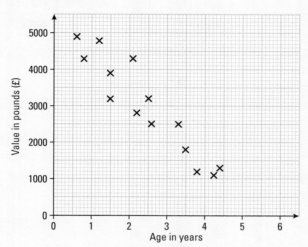

a Describe the relationship between the age of a car and its value in pounds.

b Draw a line of best fit on a copy of the scatter graph.

A car is 3 years old.

c Find an estimate of its value.

A car has a value of £3500.

d Find an estimate of its age.

March 2008

8 Jake recorded the weight, in kg, and the height, in cm, of each of ten children.
The scatter graph shows information about his results.

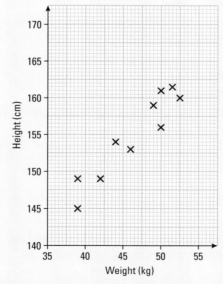

a Describe the relationship between the weight and the height of these children.

b Draw a line of best fit on a copy of the scatter graph.

c Estimate the height of a child whose weight is 47 kg.

June 2008

C

9   The table gives some information about the heights and weights of 10 athletes.

Height (cm)	180	165	185	190	178	184	168	188	192	200
Weight (kg)	73	70	80	86	75	75	72	83	85	86

a Using a horizontal scale from 160 to 210 cm and a vertical scale from 65 to 90 kg, draw a scatter graph for these data.

b Describe the correlation.

AO2 AO3

c Describe the relationship between height and weight and estimate the weight of an athlete who is 175 cm tall.

10   In a study to see how effective a weight-reducing drug was, data regarding the weight loss, in lbs, and the length of treatment, in months, was collected.
Some of the results are shown in the scatter graph.

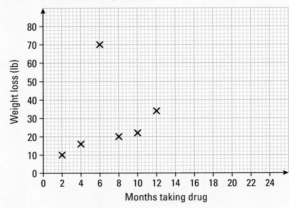

a Copy the scatter graph.

Three more people's records are taken.

Jackie lost 70 lbs in 22 months. Joan lost 60 lbs in 18 months and Tim lost 55 lbs in 14 months.

b Add these pieces of data to your scatter graph.

c One piece of data seems to be an isolated point. Circle it. Do you think this is a genuine piece of data? Give a reason for your answer.

d Ignoring the isolated point, describe the correlation.

AO3

e Write down whether or not you think the drug is effective. Give a reason for your answer.

11   The scatter diagram shows the amount of fertiliser used and the crop yields on 10 equal-size plots at a crop regulatory centre.

a Describe the correlation.

b Describe the relationship between crop yield and amount of fertiliser used.

c Estimate the crop yield when 4 kg per 80 m² of fertiliser is used.

AO3

d Estimate the amount of fertiliser used to give a crop yield of 15 000 kg.

e Nassim says he will use the line of best fit to find out what the crop would be if 20 kg of fertiliser per 80 m² was put on a plot. Will Nassim get a sensible result? Explain your answer.

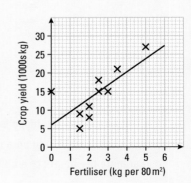

12 The numbers of fleas kept in an enclosed environment were counted every 6 days. The results are shown in the table.

Day	0	1	2	3	4	5	6
Number of fleas	50	100	196	390	780	1550	3000

  a Draw a scatter diagram of these data.

  b Draw in a curve of best fit.

  c Suggest a suitable general equation for the relationship between these data.

13 A scatter diagram is drawn to show the height above sea level (x) and air temperature (y). The equation of the line of best fit is $y = -0.01x + 19$.

  a Interpret the gradient in context.

  b What does the number 19 in the equation tell you about the height above sea level and air temperature?

Philippe Petit is a French high wire artist famous for walking along cables suspended from well-known buildings. Seen here walking across the city of Frankfurt in Germany, Petit's most celebrated and most perilous exploit was negotiating the 43m gap between the World Trade Center towers in New York, despite the probability of him surviving a fall being zero. When asked why he attempted such dangerous feats he replied "When I see three oranges, I juggle; when I see two towers, I walk."

Objectives

In this chapter you will:
- use fractions, decimals and percentages in problems
- learn how to use a number to represent a probability
- learn how to find and estimate probabilities
- learn how to add and multiply probabilities
- solve problems using ratio and proportion.

probability   fraction

5.1 Writing probabilities as numbers

Objectives

○ You can use a number to represent a probability.
○ You can add, subtract, multiply and divide fractions and decimals.
○ You can convert between fractions, decimals and percentages
○ You can use a sample space diagram to record all possible outcomes.

Why do this?

Banks use probability to assess risk when they plan takeovers of other companies.

Get Ready

Represent how likely each of these events is on a probability scale:
1. A man will jump over the Eiffel tower.
2. When you add 2 and 2 together you get the answer 4.
3. When you spin a coin you will get a head.

$$\begin{array}{c|c|c} \vdash & \vdash & \dashv \\ 0 & \frac{1}{2} & 1 \end{array}$$

Fractions

Key Points

⊚ A **probability** that lies between 0 and 1 can be expressed as a fraction or a decimal.
⊚ A **fraction** such as $\frac{3}{4}$ is made up of a **numerator** and a **denominator**.
   The denominator tells you how many parts the whole is divided into, in this case 4.
   The numerator tells you how many of these parts you have, in this case 3.
⊚ To add (or subtract) fractions, change them to **equivalent fractions** that have the same denominator.
   Then add (or subtract) the numerators but do not change the denominators.
⊚ To multiply fractions multiply the numerators together and then the denominators.
   Simplify by cancelling where possible.
⊚ To divide by a fraction turn the fraction upside down and multiply. Turn **mixed numbers** into **improper fractions** before dividing by them.
⊚ To convert a fraction into a decimal you divide the numerator by the denominator.

Example 1

Work out $1\frac{2}{3} + 4\frac{1}{2} - 2\frac{1}{8}$

$= (1 + 4 - 2) + \left(\frac{2}{3} + \frac{1}{2} - \frac{1}{8}\right)$ ← Add and subtract the whole numbers.

$= 3 + \frac{16}{24} + \frac{12}{24} - \frac{3}{24}$ ← A common denominator for the fractions is 24 (3, 2 and 8 all divide into 24). Multiply the denominator and numerator of each fraction by the same number so that the denominator becomes 24.

$= 3 + \frac{16 + 12 - 3}{24}$ ← Add and subtract the numerators.

$= 3 + \frac{25}{24}$ ← $\frac{25}{24}$ is an improper fraction.

$= 3 + 1\frac{1}{24}$ ← 24 divides into 25 once with remainder 1.

$= 4\frac{1}{24}$ ← Add together to get the answer.

numerator   denominator   equivalent fraction   mixed number   improper fraction   **117**

Example 2 Work out $2\frac{1}{2} \times 1\frac{3}{5} \div 1\frac{3}{8}$

$= \frac{5}{2} \times \frac{7}{5} \div \frac{11}{8}$ ← Convert to improper fractions. To divide turn the fraction upside down and multiply.

$= \frac{\cancel{5}^1}{{}_1\cancel{2}} \times \frac{7}{\cancel{5}_1} \times \frac{\cancel{8}^4}{11}$ ← Cancel and multiply together the numerators and denominators.

$= \frac{28}{11} = 2\frac{6}{11}$ ← Simplify to a mixed number.

Example 3 Convert $\frac{3}{8}$ into decimal.

$$\begin{array}{r} 0.375 \\ 8\overline{)3.000} \\ \underline{2\ 4} \\ 60 \\ \underline{56} \\ 40 \\ \underline{40} \end{array}$$

Another way of doing this is to divide 3 by 8 using a calculator.

Example 4 Convert 0.875 to a fraction

$0.875 = \frac{875}{1000} = \frac{35}{40} = \frac{7}{8}$

Using place value, 0.875 is 8 tenths and 7 hundredths and 5 thousandths. This is the same as 875 thousandths.

Cancel to make the fraction in its simplest form.

Percentages

Key Points

- **Percentage** (%) means number of parts per hundred.
- To write fractions, decimals and percentages in order of size first convert them all to decimals.

Example 5 Convert 35% to a fraction and a decimal.

$35\% = \frac{35}{100} = 0.35$ ←

Put the percentage over 100.
To change to a decimal drop the % sign and move the decimal point 2 places to the left.

Example 6 Use a calculator to write these in order of size. $0.23 \quad \frac{3}{8} \quad 35\% \quad \frac{1}{4}$

0.23 is already a decimal

$\frac{3}{8} = 0.375$ ← $\boxed{3} \boxed{\div} \boxed{8} \boxed{=}$

$35\% = 0.35$

$\frac{1}{4} = 0.25$ ← $\boxed{1} \boxed{\div} \boxed{4} \boxed{=}$

In order of size 0.23, 0.25, 0.35, 0.375.   Or $0.23, \frac{1}{4}, 35\%, \frac{3}{8}$

Exercise 5A

1 a Convert the following to decimals: $\frac{1}{10}$, $\frac{3}{5}$, 50%, 46%.

   b Convert the following to fractions: 0.4, 0.75, 45%, 72%.

   c Convert the following to percentages: 0.62, 0.4, $\frac{3}{10}$, $\frac{3}{20}$.

2 Write the following in order, starting with the smallest: 0.3, 48%, $\frac{1}{10}$, $\frac{3}{5}$.

3 Louise gets a mark of 28 out of 40 in a test.
  Work out what this is as a percentage.

4 Work out:

   a $1\frac{1}{2} + 4\frac{3}{4} - \frac{1}{8}$

   b $2\frac{1}{2} \times 4\frac{3}{4} \div \frac{1}{8}$

B

Key Points

- The probability P that an **event** will happen is a number in the range $0 \leqslant P \leqslant 1$.

- For an event which is **certain** $P = 1$.

- For an event which is **impossible** $P = 0$.

- A probability can be written as a fraction, a decimal or a percentage.

- For **equally likely** outcomes, the probability that an event will happen is

  $$\text{Probability} = \frac{\text{number of successful outcomes}}{\text{total number of possible outcomes}}$$

- A sample space is all the possible **outcomes** of one or more events.

- These outcomes can be presented in a **sample space diagram**.

Example 7   This 5-sided spinner is spun.

The spinner is fair.

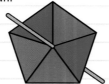

> A **fair** spinner is one that has an equal **chance** of landing on any of its sides, so each side is equally **likely**.

Work out the probability that the spinner will land on red.

> The spinner has 2 red sectors, so the number of successful outcomes = 2.

> $\frac{2}{5}$ can also be expressed as 40% or 0.4.

$p(\text{red}) = \dfrac{2}{5}$

> The spinner has 5 sides altogether, so the total number of possible outcomes = 5.

impossible   equally likely   outcome   sample space diagram   fair   chance   likely   **119**

Exercise 5B

1 Rashid spins this 7-sided spinner. The spinner is fair.
Work out the probability that the spinner will land on
a yellow
b red
c white.

2 Work out the probability of each of the following.
a rolling the number 4 with an ordinary dice
b rolling an even number with an ordinary dice
c taking an ace from an ordinary pack of cards
d taking a diamond from an ordinary pack of cards
e taking a black King from an ordinary pack of cards

3 A bag contains 5 red balls and 4 green balls. A ball is taken at random from the bag.
Work out the probability that the ball will be:
a red   b green   c yellow.

4 The faces of an 8-sided dice are numbered from 1 to 8.
Work out the probability of rolling each of the following.
a an odd number   b an even number   c a 4 or a 5
d a prime number   e a factor of 10

5 2500 tickets are sold in a school raffle. Chelsy buys 5 tickets in the raffle.
Work out the probability that she will win the raffle.

6 A box contains 3 bags of salt and vinegar crisps, 4 bags of cheese and onion crisps and 2 bags of beef crisps. One of these bags of crisps is taken from the box at random.
Work out the probability that the bag of crisps will be:
a salt and vinegar   b cheese and onion
c beef   d cheese and onion or beef.

7 A letter is chosen at random from the word PROBABILITY. Write down the probability that it will be:
a B   b Y   c R or I   d a vowel   e G.

8 The table gives the numbers of boys and the numbers of girls in a primary school and whether they are left-handed or right-handed.

	Left-handed	Right-handed	Total
Boys	47	135	
Girls	61	119	
Total			362

a Copy and complete the table.
b One of these children is chosen at random.
Use the information in your table to work out the probability that the child will be:
  i a boy
  ii left-handed
  iii a right-handed girl.

9 The pie chart gives information about how some students travelled to school one day.
One of these students is chosen at random.
Use the information in the pie chart to work out the probability that the student:

a travelled to school by bus

b walked to school.

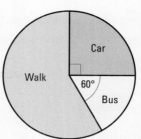

10 In a group of students
55% are boys
65% prefer to watch film A
10% are girls who prefer to watch film B.

One of these students is picked at random.
Work out the probability that the student is a boy who prefers to watch film A.

Example 8 ▷ Two ordinary dice are rolled. Work out the probability that the total score on the two dice will be 8.

	6	(1,6)	(2,6)	(3,6)	(4,6)	(5,6)	(6,6)
	5	(1,5)	(2,5)	(3,5)	(4,5)	(5,5)	(6,5)
Red	4	(1,4)	(2,4)	(3,4)	(4,4)	(5,4)	(6,4)
dice	3	(1,3)	(2,3)	(3,3)	(4,3)	(5,3)	(6,3)
	2	(1,2)	(2,2)	(3,2)	(4,2)	(5,2)	(6,2)
	1	(1,1)	(2,1)	(3,1)	(4,1)	(5,1)	(6,1)
		1	2	3	4	5	6

Blue dice

$$P(8) = \frac{\text{number of successful outcomes}}{\text{total number of possible outcomes}}$$

$$= \frac{5}{36}$$

> Draw a sample space diagram. A sample space diagram shows all the possible outcomes, for example (6, 4) represents the outcome of throwing a 6 on the blue dice and a 4 on the red dice.

> There are a total of 36 possible outcomes.

> Identify all the outcomes that give a total score of 8. There are 5 outcomes that give a total score of 8: (2, 6), (3, 5), (4, 4), (5, 3) and (6, 2).

Exercise 5C

1 Two ordinary dice are rolled.
Use the sample space diagram in Example 8 to work out the probability of getting each of the following outcomes.

a a total score of

  i 2   ii 5   iii 10 or more

b the same number on each dice

c a number on the blue dice exactly 2 more than the number on the red dice

C

2 An ordinary dice is rolled and a fair coin is spun.

   a Copy and complete the following sample space diagram to show all the possible outcomes.

Coin	H	(1,H)					
	T	(1,T)	(2,T)				
		1	2	3	4	5	6

                                Dice

   b Work out the probability of getting:

      i a 1 on the dice and a head on the coin

      ii a number greater than 3 on the dice

      iii a number less than 3 on the dice and a tail.

3 Two fair 4-sided spinners are spun and the difference between the numbers is calculated.

   a Copy and complete this sample space diagram to show all the possible outcomes.

Spinner A

Spinner B		**1**	**2**	**3**	**4**
	1	0	1		
	2	1			
	3				
	4				

   b Work out the probability of getting a difference of:

      i 0   **ii** 3   **iii** 4.

B

4 An ordinary dice is rolled and a fair 3-sided spinner is spun.

   a Draw a sample space diagram to show all the possible outcomes.

   b Use your sample space diagram to work out the probability of getting a total score of:

      i 7   **ii** 3   **iii** less than 5

5 Sunti has two boxes of crayons, A and B.

In box A he has a red crayon, a blue crayon, a yellow crayon and a black crayon.

In box B he has a red crayon, a yellow crayon and a blue crayon.

Sunti takes a crayon at random from each box.

   a Draw a sample space diagram to show all the possible outcomes.

   b Work out the probability that the crayons will be:

      i both red   **ii** the same colour   **iii** different colours.

A

6 Andy, Brigitta, Carrie, Dean and Eli are playing in a tennis competition. Each player in the competition plays every other player. There are ten matches altogether.

Two players are picked at random to play the first game.

Work out the probability that the first game will be played by a male player and a female player.

A*
A03

7 A fair 10-sided dice and an ordinary 6-sided dice are each rolled. The numbers rolled by the 10-sided dice are used for the x-coordinates, and the numbers rolled by the 6-sided dice are used for y-coordinates.

Find the probability that the point generated by the numbers on the two dice lies on each of the following lines.

   a $y = 1$ **b** $x + y = 7$ **c** $y = x + 5$ **d** $y = 2x - 5$ **e** $y = \frac{1}{2}x + 1$

5.2 Using fractions, decimals and percentages in problems

◎ Objectives

- ○ You find a percentage of a quantity.
- ○ You use decimals to find quantities.
- ○ You understand the multiplicative nature of percentages as operators.

❓ Why do this?

You have a token that gives you 20% off the price of a book. Because there is a sale on you are given another 10% off the book.
Do you get 30% off the book?

⬆ Get Ready

Convert decimals and percentages to fractions in their simplest form:

1. **a** 0.25   **b** 0.6
2. **a** 75%   **b** 35%

Finding a fraction of an amount

Key Points

- ◉ To find a fraction of an amount when the numerator of the fraction is 1, divide the amount by the denominator.
- ◉ To find a fraction of an amount when the numerator of the fraction is more than 1, divide the amount by the denominator then multiply the result by the numerator.

Example 9   Work out

     a $\frac{1}{9}$ of 62   **b** $\frac{5}{9}$ of 62

a $\frac{1}{9}$ of $62 = 62 \div 9 = 6\frac{8}{9}$ ←   *Divide by the denominator.*

b $\frac{5}{9}$ of $62 = 5 \times \frac{1}{9}$ of $62 = 5 \times 6\frac{8}{9} = 30\frac{40}{9}$ ←   *Multiply by the numerator.*

        $= 30 + 4\frac{4}{9} = 34\frac{4}{9}$

Finding a percentage of a quantity

Key Points

- ◉ To find a percentage of a quantity write the percentage as a fraction and then multiply the quantity by the fraction.

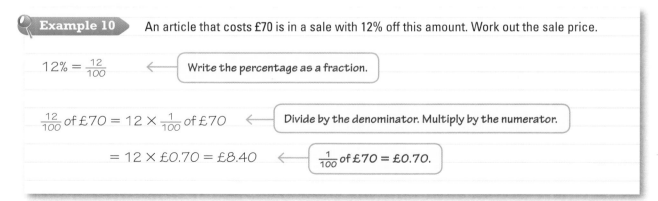

Example 10   An article that costs £70 is in a sale with 12% off this amount. Work out the sale price.

$12\% = \frac{12}{100}$ ←   *Write the percentage as a fraction.*

$\frac{12}{100}$ of £70 $= 12 \times \frac{1}{100}$ of £70   ←   *Divide by the denominator. Multiply by the numerator.*

        $= 12 \times £0.70 = £8.40$ ← $\frac{1}{100}$ *of £70 = £0.70.*

Example 11 You have a token that gives you 10% off the price of a book. Because there is a sale on you are given a further 10% off a book. The book costs £18 before deductions.
Work out the amount you would have to pay for the book.

10% of £18 $= \dfrac{£18}{100} \times 10$ ← Work out 10% of the original price.

$= £0.18 \times 10$

$= £1.80$

After 10% reduction price is £18.00 − £1.80 = £16.20 ← Take 10% from the price.

10% of £16.20 $= \dfrac{£16.20}{100} \times 10$ ← Work out 10% of the new price.

$= £0.162 \times 10$

$= £1.62$

After further 10% reduction the price is £16.20 − £1.62 = £14.58. ← Subtract from the new price to give the answer.

Finding one quantity as a percentage of another

Key Point

- To find one quantity as a percentage of another, write down the first quantity as a fraction of the second quantity then convert the fraction to a percentage.

Example 12 Kayla got 16 out of 20 in a test. Write her mark as a percentage.

16 out of 20 $= \dfrac{16}{20} = \dfrac{8}{10} = \dfrac{80}{100} = 80\%$

Compound interest

Key Points

- Banks and building societies pay **compound interest**.
- At the end of the first year, interest is paid on the money in an account. This interest is then added to the account. At the end of the second year, interest is paid on the total amount in the account, that is, the original amount of money plus the interest earned in the first year.

◉ At the end of each year, interest is paid on the total amount in the account at the start of that year.

For example, if £200 is invested in a bank account and interest is paid at a rate of 5% then

   ◉ after 1 year there will be a total of £(200 × 1.05) in the account

   ◉ after 2 years there will be a total of £((200 × 1.05) × 1.05) in the account

   ◉ after 3 years there will be a total of £(((200 × 1.05) × 1.05) × 1.05) in the account.

◉ In general, to work out the amount in a savings account after n years if interest is paid at r% per annum, multiply the original amount by $\left(1 + \dfrac{r}{100}\right)^n$.

For example to find the amount in the account after 3 years, the original £200 is multiplied by $1.05 \times 1.05 \times 1.05$ which is equivalent to 1.05^3 or 1.157 625. This is the single number that £200 is multiplied by to find the amount in the bank account after 3 years.

Example 13   £250 is invested for three years at a compound interest rate of 3% per year.

                     Work out the total after three years.

$100\% + 3\% = 103\% = 1.03$ ←   Work out the multiple.

Total after 3 years $= £250 \times (1.03)^3$ ←   Multiply £250 by the multiplier 1.03

$= £250 \times 1.092727 = £273.18$ to the nearest penny.

Depreciation

Key Points

◉ If a company buys a machine the value of the machine decreases each year by a fixed percentage. This decrease in value is known as depreciation.

◉ Depreciation can be taken to be a negative increase so that if the machine cost an amount P, and depreciates by r% at the end of each year, the value after n years is given by $P\left(1 - \dfrac{r}{100}\right)^n$.

Example 14   A machine costs £12 000 when new. It depreciates at 10% a year.

                     Work out its value after 2 years.

$$\text{Value} = £12\,000 \times (1 - 0.1)^2$$
$$= £12\,000 \times (0.9)^2$$
$$= £12\,000 \times 0.81$$
$$= £9720$$

Exercise 5D

1 Giving your answer as a mixed number work out:

   a $\frac{1}{6}$ of 67

   b $\frac{5}{6}$ of 67.

2 A radio has a price tag of £400. In the sale there is a 20% reduction. What is the sale price?

3 Mr. Hartop borrows £1500 from the bank for one year in order to buy a new washing machine.
   The bank charges him simple interest of 16% per year. How much does he have to pay back to the bank?

4 Janine wants to buy a TV that costs £280. Her mother says that she will give her a present of $\frac{1}{5}$ of the money.
   Janine has £220 herself.
   Can she afford to buy the TV or not? Give reasons for your answer.

5 A digital camera cost £220. There is 15% off in the sale. Work out the sale price.

6 Chloe buys a dress for £60 in a sale. She notices that the original price was £75.
   Work out the % reduction there was in the sale.

7 Jack puts £600 into a three-year savings account that gives compound interest of 4% per year.
   Work out how much Jack will have at the end of three years.
   Give your answer to the nearest penny.

* 8 Mr Jackson has a hardware store card that gives him 10% off all purchases.
   The store has a sale on that gives a further 12% reduction after the 10% has been taken off.
   Mr Jackson buys a stainless steel spade that costs £25 before deductions.
   The salesman says that the spade will cost him £19.50.
   Mr Jackson thinks that the salesman is not charging the correct amount.
   Discuss, with reasons, whether or not Mr Jackson is right.

* 9 Sophie wants to invest £800 for two years.
   She wants to get as good a rate of return as possible.
   The building society offers her two accounts.
   The first gives 3.5% interest for year one and then 2.5% interest for year two.
   The second gives compound interest of 3% per year.
   Decide, with reasons, which account you think that Sophie should use.

* 10 James buys a new car that costs £9500.
   In the first year it depreciates by 45%.
   In years two and three it depreciates by 20% each year.
   At the end of year three James advertises his car for sale at £3000.
   Decide, with reasons, whether or not this is a fair price.

5.3 Mutually exclusive outcomes

◉ Objectives

- You can add probabilities for mutually exclusive events.
- You can work out the probability that something will not happen given that you know the probability that it will happen.

❓ Why do this?

If you are planning a barbeque and you know that there is a 1 in 3 chance of it raining on that day, you can work out the probability that it won't rain.

◈ Get Ready

Work out the missing numbers.

a $0.3 + ? = 0.7$ b $1 - ? = 0.35$ c $\frac{2}{5} + ? = 1$ d $? + \frac{1}{3} = \frac{1}{2}$

🔑 Key Points

- Two events are **mutually exclusive** when they cannot occur at the same time.
- For mutually exclusive events A and B,
  P(A or B) = P(A) + P(B).
- For 3 or more events,
  P(A or B or C or ...) = P(A) + P(B) + P(C) + ...
- For mutually exclusive events A and not A,
  P(not A) = 1 − P(A).
- If a set of mutually exclusive events contains all possible outcomes, then the sum of their probabilities must come to 1. $\Sigma p = 1$.
- For three mutually exclusive events that cover all possible outcomes, P(A) + P(B) + P(C) = 1.

🔍 Example 15

A card is taken at **random** from an ordinary pack of cards. Work out the probability that the card will be an ace or the 10 of clubs.

> Use $P(A) = \dfrac{\text{number of successful outcomes}}{\text{total number of possible outcomes}}$

$P(\text{ace or } 10\clubsuit) = P(\text{ace}) + P(10\clubsuit)$

$\qquad = \dfrac{4}{52} + \dfrac{1}{52}$

$\qquad = \dfrac{5}{52}$

> There are 52 cards in an ordinary pack of cards, so the total number of possible outcomes = 52. There are four ways to pick an ace, so there are four successful outcomes A♣, A♦, A♥, A♠. So $P(\text{ace}) = \dfrac{4}{52}$.

> There is only one 10 of clubs, so $P(10\clubsuit) = \dfrac{1}{52}$.

⚙ Exercise 5E

1. Here is a number of shapes.

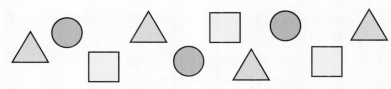

One of these shapes is chosen at random. Work out the probability that the shape will be:

a a square b a triangle c a square or a triangle.

D

2   A bag contains some counters. The colour of each counter is either red, or black, or white. A counter is taken at random from the bag. The probability that the counter will be red is 0.3. The probability that the counter will be white is 0.6. Work out the probability that the counter will be red or white.

3   The table gives the probability of getting each of 1, 2, 3 and 4 on a biased 4-sided spinner.

Number	1	2	3	4
Probability	0.2	0.35	0.15	0.3

Work out the probability of getting:

a   1 or 4   b   2 or 3   c   2 or 4   d   1 or 2 or 3

4   Below is a number of lettered tiles.

One of these tiles is selected at random.
Work out the probability of getting:

a   an A   b   an L   c   an O

d   an A or an L   e   an A or an O   f   an L or an O.

5   Below are some cards with coloured letters.

X	Z	X	X	Y	X
Y	Z	Z	Y	Z	Y
Z	X	X	Y	X	Z

One of these cards is picked at random. Work out the probability that the letter on the card will be:

a   red   b   an X   c   a red X   d   red or an X.

C

6   A card is taken at random from an ordinary pack of cards.
Work out the probability of getting:

a   a 3 of hearts or a 5 of spades

b   a heart or a spade

c   a King of clubs or a Queen of any suit

d   a diamond or the ace of hearts

e   a picture card (Jack, Queen or King) or a red 10.

B

7   A and B are two mutually exclusive events. P(A) = 0.45 and P(A or B) = 0.8.
Work out the value of P(B).

A

* 8   Paul rolls an ordinary dice. He says that the probability of getting a 6 on the dice is $\frac{1}{6}$, and the probability of getting another 6 when the dice is rolled again is $\frac{1}{6}$, so the probability of getting two sixes is $\frac{1}{6} + \frac{1}{6} = \frac{2}{6} = \frac{1}{3}$. Is he correct? Explain why.

Example 16

A bag contains 10 balls. Three of the balls are green. A ball is taken at random from the bag. Work out the probability that the ball will be:

a green

b not green.

> Use $P(A) = \dfrac{\text{number of successful outcomes}}{\text{total number of possible outcomes}}$

> There are 10 balls altogether, so the total number of possible outcomes = 10.

a $P(\text{green}) = \dfrac{3}{10}$ ← 3 of these outcomes result in successfully taking a green ball, so $P(\text{green}) = \dfrac{3}{10}$

b Using $P(\text{not } A) = 1 - P(A)$,

$P(\text{not green}) = 1 - P(\text{green})$

$= 1 - \dfrac{3}{10}$

$= \dfrac{7}{10}$

> Subtract $\dfrac{3}{10}$ from 1.
>
> $1 - \dfrac{3}{10} = \dfrac{10}{10} - \dfrac{3}{10}$
>
> $= \dfrac{10 - 3}{10} = \dfrac{7}{10}$

Exercise 5F

1   The pie chart shows the proportions of people voting Labour (L), Conservative (C) and Liberal Democrat (LD) in a town.

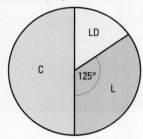

One of these voters is chosen at random for an opinion poll.
Work out the probability that the person voted Liberal Democrat.

2   The probability of rolling a 5 on a biased dice is $\frac{1}{5}$.
Work out the probability of not rolling a 5.

3   The probability that it will rain tomorrow is 0.65.
Work out the probability that it will not rain tomorrow.

4   Harry has a 75% chance of hitting a treble 20 with a dart.
Work out the probability that Harry will not hit the treble 20 with a dart.

5   A card is taken at random from an ordinary pack of cards.
Work out the probability that the card will be:
a   an ace
b   not an ace.

D

C

C

6 The sectors of a 3-sided spinner are coloured brown, blue and black.
The table gives information about the probability of getting brown, and blue, on the spinner.

Colour	Brown	Blue	Black
Probability	0.35	0.15	

The spinner is spun.
a Work out the probability of getting a colour that is:
  i not brown   ii not blue.
b Jamie says the probability of getting black with this spinner is 0.5. He is right. Explain why.

B

7 A bag contains red balls, blue balls and green balls in the ratio 2:3:4.
A ball is taken at random from the bag. Work out the probability that the ball will be:
a red   b not red   c not green.

A
A03

8 For two mutually exclusive events A and B, P(B) = 0.3 and P(A or B) = 0.7.
Work out P(not A).

5.4 Estimating probability from relative frequency

◎ Objective

● You can find an estimate for a probability from the results of an experiment.

？ Why do this?

If you tossed a coin a large number of times you could calculate the probability of getting a head. If this isn't 0.5 then it might show that the coin is biased.

◈ Get Ready

Write these fractions in their simplest form.

a $\frac{10}{20}$ b $\frac{12}{16}$ c $\frac{72}{81}$ d $\frac{39}{169}$

◉ Key Points

◉ You can use **relative frequency** to find an estimate of a probability.

Estimated probability $= \dfrac{\text{number of successful trials}}{\text{total number of trials}}$

◉ The greater the number of **trials**, the more accurate the estimated probability.

Example 17   Suki rolls a dice 120 times. Here are the results of her experiment.

Number	1	2	3	4	5	6
Frequency	19	18	23	22	21	17

Work out an estimate for the probability of rolling a 6 on Suki's dice.

Estimated probability $= \dfrac{17}{120}$ ⟵ From the table, the number of trials which successfully result in rolling a 6 = 17, and the total number of trials = 120.

Exercise 5G

1. Tania spins a coin 100 times and gets 45 heads.
   Work out an estimate for the probability of getting a head on Tania's coin.

2. A gardener plants 60 seeds. 52 of these seeds germinate.
   Work out an estimate for the probability that this type of seed will germinate.

3. The sectors of a 3-sided spinner are each coloured red or orange or green.
   The table gives the results when the spinner is spun 300 times.

Colour	Red	Orange	Green
Frequency	154	56	90

   a Use the information in the table to find an estimate for getting red.

   b Is this a fair spinner? Give a reason for your answer.

4. Drop a drawing pin 50 times and record whether it lands on its head or on its tail.

   a Use your results to find an estimate for the probability of the drawing pin landing on its head.

   b How could you improve on your answer to part **a**?

   Head   Tail

5. Domenique records the numbers of 6s she gets when she rolls a dice 10, 100 and 1000 times.
   The table below shows her results.

Number of rolls	10	100	1000
Number of 6s	1	15	165

   Use this information to work out the best estimate for getting a 6 on Domenique's dice.
   Give a reason for your answer.

* 6. Malik says that when he drops a piece of toast it always lands butter-side down.
   Carry out an experiment to find an estimate for the probability that a piece of toast will land butter-side down.
   Explain all stages of your work.

5.5 Finding the expected number of outcomes

◎ Objective

● You can find the expected number of outcomes in an experiment.

⑦ Why do this?

Knowing the probability that a small sample of people, such as your class, have a pet means that you can estimate the number of people in your school who have a pet.

◈ Get Ready

Work out the following.

a $\frac{1}{2} \times 100$ **b** $\frac{1}{3} \times 54$ **c** $\frac{2}{5} \times 75$ **d** $\frac{3}{8} \times 108$

Example 18 The probability of winning a prize in a raffle is $\frac{1}{25}$. Jaqui buys 100 tickets in the raffle. How many prizes can she expect to win?

The probability that Jaqui will win a prize is $\frac{1}{25}$.
So she can expect to win 1 prize in every 25 tickets she buys.

> Expected number of outcomes
> = Number of trials × Probability.

Jaqui buys 100 tickets so she can expect
to win $\frac{1}{25} \times 100 = 4$ prizes.

⚙ Exercise 5H

1 Lauren spins an ordinary coin 100 times. How many heads can she expect to get?

2 Yousif rolls an ordinary dice 60 times. How many 6s can he expect to get?

3 The table gives the probability of spinning the numbers 1, 2, 3 and 4 on a 4-sided spinner.

Number	1	2	3	4
Probability	0.2	0.35	0.15	0.3

Vicky spins the spinner 200 times. Work out an estimate for the number of 3s that Vicky will get.

4 A bag contains 1 red peg, 5 white pegs and 4 yellow pegs.
A peg is taken at random from the bag and then replaced. This is done 250 times.
Copy and complete the table to show the expected numbers of red, white and yellow pegs that will be taken from the bag.

Colour of peg	Red	White	Yellow
Expected number			

5 Fatima spins two coins and records the result. She does this 120 times. One possible outcome is (head, head). Find an estimate for the number of times she will get two heads.

* 6 The probability of winning a prize in a lottery is $\frac{1}{50}$. Austin says that if he buys 50 tickets in the lottery he will win a prize. Is he right? Give a reason for your answer.

7 A card is taken from an ordinary pack of cards. It is then replaced. If this is done 260 times, work out the expected number of the following that will be taken from the pack.

    a Jacks

    b spades

    c aces or clubs

* **8** A doctor estimates that the probability a patient will come to see her about a bad back is 0.125.
Of the next 240 patients who come to see her, 20 have a bad back.
How good is the doctor's estimate of this probability? Explain your answer.

A

5.6 Independent events

◉ Objectives

○ You can find the probability of independent events.
○ You can multiply probabilities.

◈ Why do this?

If you pick one sweet from a bag and replace it because you don't like the flavour, then you pick again, you have exactly the same probability of choosing the same flavour again.

◈ Get Ready

Work out the following. Give your answers in their simplest form.

a $\frac{1}{2} \times \frac{1}{3}$ **b** $\frac{1}{3} \times \frac{6}{7}$ **c** $\frac{2}{3} \times \frac{3}{5}$ **d** $\frac{5}{12} \times \frac{9}{20}$

◈ Key Points

◉ Two events are independent if one event does not affect the other event.
◉ For two **independent events** A and B,
    $P(A \text{ and } B) = P(A) \times P(B)$
◉ For 3 or more events,
    $P(A \text{ and } B \text{ and } C \text{ and } \ldots) = P(A) \times P(B) \times P(C) \times \ldots$

Example 19   An ordinary coin is spun and an ordinary dice is rolled.
         Work out the probability of getting a head on the coin and a 6 on the dice.

$P(\text{head and } 6) = P(\text{head}) \times P(6)$ ←

> The event of getting a head on the coin cannot affect the event of getting a 6 on the dice. They are independent events. Use $P(A \text{ and } B) = P(A) \times P(B)$. Here A = head and B = 6.

$= \frac{1}{2} \times \frac{1}{6}$ ←   Multiply the fractions.

$= \frac{1}{12}$ ←   So $\frac{1}{2} \times \frac{1}{6} = \frac{1 \times 1}{2 \times 6} = \frac{1}{12}$

Exercise 5I

C

1. Dan and Nicole play a game of chess and a game of draughts.
    The probability that Dan will win the game of chess is 0.4.
    The probability that Dan will win the game of draughts is 0.8.
    Work out the probability that Dan will win both games.

2. The probability that it will rain tomorrow is $\frac{2}{3}$.
    The probability that Jaleel will forget his umbrella tomorrow is $\frac{3}{4}$.
    Work out the probability that it will rain tomorrow and Jaleel will forget his umbrella.

3. The probability that a postman will deliver mail to Yasmin's house tomorrow is 0.8.
    The probability that Yasmin's dog will bark when the postman delivers the mail is 0.75.
    Work out the probability that a postman will deliver mail to Yasmin's house tomorrow and her dog will bark.

4. The probability that Joshua will forget his protractor for an examination is 0.35.
    The probability that he will forget his calculator for the examination is 0.15.
    Work out the probability that he will:
    a not forget his protractor
    b not forget his calculator
    c not forget his protractor and not forget his calculator.

B

5. A card is taken at random from each of two ordinary packs of cards, pack A and pack B.
    Work out the probability of getting:
    a a red card from pack A and a red card from pack B
    b a diamond from pack A and a club from pack B
    c a King from pack A and a picture card (King, Queen, Jack) from pack B
    d a 10 from pack A and a 10 of clubs from pack B
    e an ace of hearts from each pack.

A

6. Tony and Hannah each have some coins. The table below gives information about these coins.

	Number of coins			
	1p	2p	5p	20p
Tony	5	2	3	1
Hannah	3	1	2	2

Tony and Hannah each pick one of their own coins at random.
Work out the probability that:
a Tony picks a 5p coin and Hannah picks a 5p coin
b they both pick a silver coin
c Tony does not pick a 20p coin and Hannah does not pick a 20p coin.

5.7 Probability tree diagrams

◎ Objective

○ You can use a tree diagram to work out probabilities.

◈ Why do this?

Probability trees are used differently to family trees but they both show all the outcomes from a single starting point.

◈ Get Ready

Work out.

a $\frac{1}{2} + \frac{1}{4}$ **b** $\frac{1}{3} + \frac{1}{5}$ **c** $\frac{2}{5} + \frac{1}{4}$ **d** $\frac{3}{7} + \frac{4}{9}$

● Key Point

◉ A **probability tree diagram** shows all possible outcomes of an experiment.

Example 20

Box A contains 3 red balls and 4 blue balls.
Box B contains 2 red balls and 3 blue balls.
One ball is taken at random from each box.

  a Draw a tree diagram to show all the outcomes.
  b Work out the probability that the balls will have the same colour.

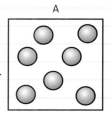
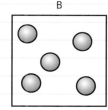

A   B

a

> Work out the probability of getting each event and show these on the tree diagram. For example, the probability of getting a ball from box A $= \dfrac{\text{number of successful outcomes}}{\text{total number of possible outcomes}} = \dfrac{3}{7}$.

Box A	Box B	Outcome	Probability
	$\frac{2}{5}$ red	red, red	$\frac{3}{7} \times \frac{2}{5} = \frac{6}{35}$
$\frac{3}{7}$ red	$\frac{3}{5}$ blue	red, blue	$\frac{3}{7} \times \frac{3}{5} = \frac{9}{35}$
$\frac{4}{7}$ blue	$\frac{2}{5}$ red	blue, red	$\frac{4}{7} \times \frac{2}{5} = \frac{8}{35}$
	$\frac{3}{5}$ blue	blue, blue	$\frac{4}{7} \times \frac{3}{5} = \frac{12}{35}$

> A tree diagram shows all the possible outcome of an experiment. For example, this branch represents taking a red ball from box A followed by a red ball from box B.

> Taking a red ball from box A and taking a red ball from box B are independent events.
> Use P(A and B) = P(A) × P(B). Here A = a red ball from box A, and B = a red ball from box B.
> So P(red and red) = P(red) × P(red) $= \frac{3}{7} \times \frac{2}{5}$.
> Multiply the fractions:
> $\frac{3}{7} \times \frac{2}{5} = \frac{3 \times 2}{7 \times 5} = \frac{6}{35}$

b P(same colour) = P(both red or both blue)

> The events are mutually exclusive; you cannot take two red balls and two blue balls from the box at the same time.

= P(red, red) + P(blue, blue)

> Use P(A or B) = P(A) + P(B). Here A = both red (i.e. red, red) and B = both blue (i.e. blue, blue).

$= \dfrac{6}{35} + \dfrac{12}{35}$

> From the tree diagram,
> P(red, red) $= \dfrac{6}{35}$ and P(blue, blue) $= \dfrac{12}{35}$

$= \dfrac{18}{35}$

> Add the fractions.

A

⚙ **Exercise 5J**

1   Bag A contains 2 blue counters and 3 white counters.

Bag B contains 3 blue counters and 4 white counters.

A counter is taken at random from each bag.

a   Copy and complete the tree diagram to show all the possible outcomes.

b   Work out the probability that the counters will both be:

   i   white

   ii   blue

   iii   the same colour.

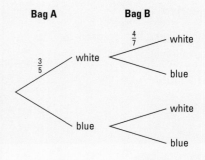

2   There are 10 pencils in a pencil case.

3 of the pencils are HB pencils.

A pencil is taken at random from the pencil case and then returned.

A second pencil is now taken at random from the pencil case and then returned.

a   Copy and complete the tree diagram to show all the possible outcomes.

b   Work out the probability that only one of the pencils will be an HB pencil.

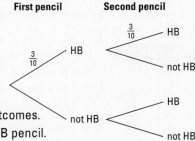

3   Ryan and Ibrahim each have a bag of sweets. In Ryan's bag there are 3 orange sweets and 5 red sweets.

In Ibrahim's bag there are 2 orange sweets and 3 red sweets.

The boys each take a sweet at random from their own bag.

a   Draw a tree diagram to show all the possible outcomes.

b   Use your tree diagram to work out the probability that the sweets will:

   i   both be orange

   ii   each have a different colour.

4   On her way home from work Taylor must drive through two sets of traffic lights.

The probability that she will be stopped at the first set of traffic lights is 0.4.

The probability that she will be stopped at the second set of traffic lights is 0.7.

a   Work out the probability that she will not be stopped at:

   i   the first set of traffic lights

   ii   the second set of traffic lights.

b   Draw a tree diagram to show all the possible outcomes.

c   Work out the probability that she will be stopped by:

   i   both sets of traffic lights

   ii   only one set of traffic lights

   iii   at least one set of traffic lights.

5   Will spins two spinners, A and B. The probability of getting a 6 on spinner A is 0.3.

The probability of getting a 6 on spinner B is 0.45.

a   Draw a tree diagram to show all the possible outcomes.

b   Work out the probability of getting a 6 on:

   i   neither spinner

   ii   only one spinner

   iii   spinner B only.

6 The probability that Steph will be late for school on Monday is $\frac{1}{4}$. The probability that Steph will be late for school on Tuesday is $\frac{2}{9}$. Work out the probability that she will be late on at least one of these days.

A02
A03 A

7 A card is taken at random from an ordinary pack of cards. It is then replaced.
Another card is now taken at random from the pack of cards.
Work out the probability of the following.
  a Both cards are Kings.
  b Neither of the cards is a heart.
  c One of the cards is a spade.
  d One of the cards is the ace of clubs.
  e At least one of the cards is a diamond.

8 Three ordinary coins are spun.
  a Show all the possible outcomes.
  b Work out the probability of getting:
    i 3 heads
    ii 2 heads and 1 tail (in any order).

A02

9 The probability that an egg has a double yolk is 0.1.
Nick has three eggs. Work out the probability that exactly one of his eggs has a double yolk.

A02

10 There is a 95% chance that a Gleemo light bulb is faulty. A shop sells Gleemo light bulbs in packets of three. The shop has 400 packets of Gleemo light bulbs in stock.
Find an estimate for the number of packets that will have exactly 2 faulty light bulbs.

A03 A*

5.8 Conditional probability

◉ Objective

○ You can find the probability of events that are not independent.

? Why do this?

If you pick two socks from a drawer, you can work out the probability that they would match using conditional probability.

⬆ Get Ready

There are 10 buttons in a box. Four of these buttons are red. Tony takes a red button from the box and sews it on his shirt. He now takes at random another button from the box. What is the probability that this button will be red?

🔵 Key Point

◉ A **conditional probability** is when one outcome affects another outcome, so that the probability of the second outcome depends on what has already happened in the first outcome.

Example 21 A bag contains 5 counters. 3 counters are white and 2 counters are black.
Two counters are taken at random from the bag.
  a Draw a tree diagram to show all the possible outcomes.
  b Find the probability that:
    i both counters will be black
    ii only one counter will be white.

a

First counter	Second counter	Outcome	Probability
		W, W	$\frac{3}{5} \times \frac{2}{4} = \frac{6}{20}$
		W, B	$\frac{3}{5} \times \frac{2}{4} = \frac{6}{20}$
		B, W	$\frac{2}{5} \times \frac{3}{4} = \frac{6}{20}$
		B, B	$\frac{2}{5} \times \frac{1}{4} = \frac{2}{20}$

Tree: First counter W ($\frac{3}{5}$) and B ($\frac{2}{5}$); from W: W ($\frac{2}{4}$), B ($\frac{2}{4}$); from B: W ($\frac{3}{4}$), B ($\frac{1}{4}$).

> Taking two counters from the bag is the same as taking one counter from the bag followed by taking another counter from the bag (the first counter is not put back before the second counter is taken).

> If the first counter is white, then the second counter will be taken from a bag containing 2 white counters and 2 black counters.
> If the first counter is black, then the second counter will be taken from a bag containing 3 white counters and 1 black counter.

ResultsPlus
Examiner's Tip

It is not necessary to simplify a probability fraction in the examination.

From the tree diagram:

b   i $P(B, B) = \frac{2}{5} \times \frac{1}{4} = \frac{2}{20}$

   ii $P(W, B \text{ or } B, W) = P(W, B) + P(B, W)$ ← Only one counter is white, i.e. either the first counter is white and the second counter is black or the first counter is black and the second counter is white.
   $= \frac{3}{5} \times \frac{2}{4} + \frac{2}{5} \times \frac{3}{4}$
   $= \frac{6}{20} + \frac{6}{20}$
   $= \frac{12}{20} = \frac{3}{5}$

Exercise 5K

A☆

1 A box contains 7 black balls and 3 white balls.
A ball is taken at random from the box and it is not replaced.
A second ball is now taken at random from the box.
  a Copy and complete the tree diagram to show all the possible outcomes.
  b Work out the probability that the balls will be:
    i black
    ii the same colour
    iii one of each colour.

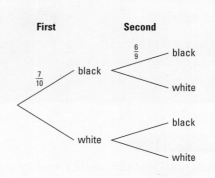

First   Second

$\frac{7}{10}$ black ⟨ $\frac{6}{9}$ black / white ⟩

white ⟨ black / white ⟩

2   5 boys and 7 girls want to be chosen for a school council.
Two of these students are picked at random.

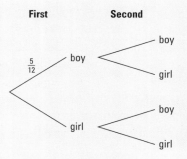

a   Copy and complete the tree diagram to show all the
possible outcomes.

b   Work out the probability of getting:
   i   2 boys
   ii 1 boy
   iii 1 or more boys.

3   Eight cards numbered 1 to 8 are shuffled thoroughly. The top two cards are turned face up on a table.
Draw a probability tree diagram and use it to work out the probability that the numbers will:

a   both be even

b   add up to an odd number.

4   On any school day, the probability that Josh oversleeps is $\frac{1}{5}$. If he oversleeps, the probability that he will
remember all his books is $\frac{2}{9}$. If he does not oversleep, the probability that he will remember all his books
is $\frac{5}{7}$. Use a tree diagram to work out the probability that Josh will not remember all of his books tomorrow.

5   Alexis travels to school by bus or by train. The probability that she travels by bus is 0.45.
If she travels to school by bus, the probability that she will be late is 0.15. If she travels to school by
train, the probability that she will be late is 0.35. Work out the probability that she will not be late.

6   Widgets Electronic Company makes two types of microprocessors, X and Y, in equal numbers. The
probability that type X will be faulty is 0.25. The probability that type Y will be faulty is 0.15. If a type X
microprocessor is faulty, the probability that it will be recycled is 0.85. If a type Y microprocessor is
faulty, the probability that it will be recycled is 0.35. The company only recycles faulty microprocessors.
A microprocessor is picked at random. Work out the probability that it will be recycled. Give your
answer in standard form.

A02
A03

7   Lizzie has seven coins in her purse. Three of the coins are 1 euro coins and four of the coins are £1
coins.
Lizzie drops her purse and two coins fall out. Work out the probability that the coins will be:

a   both £1 coins

b   not 1 euro coins.

A02

8   Two cards are taken at random from an ordinary pack of cards.
Work out the probability that the cards will be:

a   both aces

b   both hearts

c   from different suits.

A02

9   The probability that it will rain today is $\frac{3}{7}$. If it does not rain today the probability that it will rain tomorrow
is $\frac{4}{7}$. Work out the probability that it will rain today or tomorrow.

A02
A03

10 A doctor diagnoses that a patient has a virus. She does not know which type of virus, X, Y or Z, the
patient has. The probability that the patient will have a virus of type X, or Y, or Z is 0.56, or 0.28, or 0.16,
respectively. The probability that the patient will not recover from each type of virus is 0.18, 0.22 or 0.35,
respectively. Work out the probability that the patient will recover from the virus.

A02
A03

A*

5.9 Ratio and proportion

⊚ Objectives

- ○ You use ratios.
- ○ You write ratios in their simplest form.
- ○ You divide a quantity in a given ratio.
- ○ You solve a ratio problem in context.

❓ Why do this?

You want to mix petrol and oil to make a mixture suitable for a two-stroke engine. The amount of petrol to oil is given as a ratio.

⬥ Get Ready

1. Write $\frac{3}{12}$ in its simplest form.
2. Write $\frac{40}{100}$ in its simplest form.

🔧 Key Points

- ⊚ **Ratio** is the quantity of one thing compared to the quantity of something else.
- ⊚ A ratio of 1 part oil to 20 parts of petrol would be written as 1 : 20.
- ⊚ If you borrow money using a credit card you have to pay interest on the money you borrow.
  The more money you borrow the more interest you have to pay.
  If you borrow £10 and you pay £0.60 in interest then if you borrow £20 you would pay twice as much i.e. £1.20.
  The interest you pay is proportional to the amount you borrow.
- ⊚ Ratio and proportion are sometimes used to collect data.

Example 22 ▶ For every eight red cars sold by a garage they sell six silver cars and four blue ones.

Write down the ratios of the number of red cars to the number of silver to the number of blue cars.

The ratio is 8 : 6 : 4   ⟵ [8, 6 and 4 are divisible by 2.]

= 4 : 3 : 2 in the simplest form. ⟵

[4, 3 and 2 are not divisible by any other number so this is the simplest form.]

Example 23 ▶ The ratio of games won, lost and drawn by a team was 4 : 2 : 3

What fraction of the games did the team lose?

4 + 2 + 3 = 9 games in total ⟵ [For every 9 games 4 are won, 2 are lost and 3 are drawn.]

$\frac{2}{9}$ of the games were lost. ⟵

[2 out of every 9 games are lost.]

Example 24

A stainless steel contains steel, chromium and nickel in the ratio 76 : 16 : 8.
Work out how many kilograms of steel, nickel and chromium are contained in 1500 kilograms of stainless steel.

$76 + 16 + 8 = 100$

$\dfrac{1500}{100} = 15\,kg$

$15\,kg \times 76 = 1140\,kg$ so there are 1140 kg of steel.
$15\,kg \times 16 = 240\,kg$, so there are 240 kg of chromium.
$15\,kg \times 8 = 120\,kg$, so there are 120 kg of nickel.

Example 25

A spring under a load of 3 kg compresses by 4 mm. The amount the spring compresses is proportional to the load put on it. How much would the spring compress under a load of 5 kg?

$\dfrac{4mm}{3\,kg} = 1\tfrac{1}{3}\,mm/kg$ ⟵ | Find the compression for 1 kg. |

$5\,kg \times 1\tfrac{1}{3}\,mm/kg = 5 + \tfrac{5}{3} = 6\tfrac{2}{3}\,mm$ ⟵ | Multiply the compression for 1 kg by 5 kg. |

Example 26

A scientist wants to estimate the number of birds in a large aviary. He catches 20 birds and tags them. The birds are returned to the aviary. The next day the scientist catches another 20 birds. Four of these birds have tags on. Work out an estimate for the number of birds in the aviary.

| Assume the proportion of tagged birds in the second sample will be in the same proportion as in the aviary. |

$4 : 20 = 20 : n$

so $\dfrac{4}{20} = \dfrac{20}{n}$ ⟵ | Let n be the number in the aviary. |

$n = \dfrac{20 \times 20}{4} = 100$

thus $4 : 20 = 20 : 100$
There are 100 birds in the aviary.

Exercise 5L

1 Mercury Rangers played 28 matches. They won 10, lost 12 and drew the rest. Write the matches won, lost and drawn as a ratio.

2 Here is a list of the ingredients to make 30 shortbread biscuits.
220 g butter, 100 g sugar, 350 g flour
Harry is going to make 120 shortbread biscuits for a fete. How much of each ingredient will he need?

3 In a car park there are 40 white cars, 35 red cars and 55 silver cars.
   a Write down the ratio of the number of white cars to red cars to silver cars in its simplest form.
   b One car is picked at random. Work out the probability that it is silver.

4 In a factory there are 176 shop floor workers. The ratio of supervisors to shop floor workers is 1 : 15. Write down the number of supervisors.

5 In a bag of sweets the ratio of orange to yellow to white sweets is 3 : 4 : 5.
   a What fraction of the sweets is white?
   Olivia picks a sweet out of the bag.
   b What is the probability that it is an orange sweet?

6 Lynsey is 20 years old. Isla is 15 years old. They are to share £700 in the ratio of their ages. Work out how much each will get.

7 A metal is made of copper and zinc in the ratio 56 : 24.
Work out how many kilograms of copper and zinc will be in 1000 kg of the metal.

8 Katie borrows £1000 for 1 year at an agreed total interest of £95.
   a If she decides to borrow £3000 instead how much interest will she pay?
   b If she decides to borrow £550 instead how much interest will she pay?

9 A spring under a load of 2 kg compresses by 3.5 mm. The amount the spring compresses is proportional to the load put on it.
   a Work out how much the spring would compress under a load of 4 kg.
   b Do you think it would be reasonable to work out the compression under a load of 50 kg using proportionality? Give a reason for your answer.

Chapter review

● The **probability** P that an event will happen is a number in the range $0 \leqslant P \leqslant 1$.
● For an event which is **certain** $P = 1$.
● For an event which is **impossible** $P = 0$.
● A probability can be written as a fraction, a decimal or a **percentage**.
● For equally likely **outcomes**, the probability that an event will happen is

$$\text{Probability} = \frac{\text{number of successful outcomes}}{\text{total number of possible outcomes}}$$

● A sample space is all the possible outcomes of one or more events.
● These outcomes can be presented in a **sample space diagram**.

◉ Two events are **mutually exclusive** when they cannot occur at the same time.

◉ For mutually exclusive events A and B,
P(A or B) = P(A) + P(B)

◉ For mutually exclusive events A and not A,
P(not A) = 1 − P(A)

◉ You can use **relative frequency** to find an estimate of a probability.
Estimated probability = $\dfrac{\text{number of successful trials}}{\text{total number of trials}}$

◉ Two events are independent if one event does not affect the other event.

◉ For two **independent events** A and B,
P(A and B) = P(A) × P(B)

◉ A **probability tree diagram** shows all possible outcomes of an experiment.

◉ A **conditional probability** is when one outcome affects another outcome, so that the probability of the second outcome depends on what has already happened in the first outcome.

Review exercise

1   Part of the label on a packet of Weetabix is shown.

   a What fraction of the 100 g is:

      i fat   **ii** carbohydrate   **iii** fibre   **iv** protein

   b Write each of these fractions as a percentage.

   c Write each of these fractions as a decimal.

   d The total of the percentages found in **b** do not add up to 100%.
      Suggest a reason for this.

Weetabix	
Per 100g	
Fat	2g
Carbohydrate	68g
Fibre	10g
Protein	12g

A02
A03

2   A letter is picked at random from the word MISSISSIPPI.
Work out the probability that the letter will be:

   a an S   **b** an I   **c** not an S

   d not an I   **e** an S or an I   **f** neither an S nor an I.

D

3   Some students are asked which topic from algebra, geometry and statistics they like best.
The results are given in the table below.

	Algebra	Geometry	Statistics
Boys	14	14	16
Girls	9	13	24

One of these students is picked at random. Work out the probability that the student:

   a is a girl

   b likes geometry best

   c is a boy who likes statistics best

   d is a girl who does not like algebra best

   e is a boy who does not like algebra or geometry best.

4   A jar of sweets contains toffees, truffles and creams in the ration 3 : 7 : 8.
A sweet is chosen at random. Write down the probability that it is a truffle.

D

5 Sam wants to buy a car.
He decides to borrow £2500 from his father.
He adds interest of 4.5% to the loan and this total is the amount he must repay his father.
How much will Sam pay back in total?

6 Maggie, John and Mark are to receive an inheritance of £5600.
The money is to be shared out in the ratio 1 : 3 : 4. Work out how much each will get.

A02
A03

7 Strawberry ice cream uses the ingredients strawberry, cream, sugar and milk in the ratio 11 : 6 : 4 : 3.
Jolana is going to make strawberry ice cream for the summer fete.
Each portion will have 240 g of ice cream.
She has plenty of cream, milk and sugar in her store cupboard.
She picks 1.8 kilograms of strawberries.

a Work out how many full portions of ice cream Jolana can make.
At the fete Leo buys 3 portions of the strawberry ice cream for £3.66.
Lily buys 5 portions of ice cream. There is a 10% discount if you buy 5 or more portions.

b Work out how much Lily pays for the ice cream.

C

8 Mike rolls an ordinary dice and spins a fair 4-sided spinner.
By drawing a sample space diagram or otherwise, work out the
probability that the total score will be:

a 7
b less than 5
c a prime number.

A02
A03

9 Mark wants to invest £5000 in a five-year fixed interest building society account.
There are two possible accounts at his local building society.

ACCOUNT 1	ACCOUNT 2
4% Simple Interest per annum	3.5%
Paid yearly by cheque	Compound Interest
(Not added to account)	(Payable at end of 5 years)

Investigate the returns on these accounts. What advice would you give Mark?

B

10 A card is taken at random from an ordinary pack of cards.
It is then replaced. This is done 390 times.
How many times would you expect to see:

a a heart
b the ace of spades
c a Jack?

ResultsPlus
Exam Question Report

71% of students answered this sort of question
poorly because they chose the wrong approach
to calculating the probability.

* 11 Amy has a dice. She thinks it is biased. Describe what Amy could do to see if her dice is biased. Give as much detail as you can.

12 Chris spins three coins and records the results. He does this 240 times. One possible outcome is (head, head, head). Find an estimate for the number of times he will get 2 heads and 1 tail (in any order).

13 Megan buys 10 tickets in a raffle. Three of these tickets win a prize.
She says that the probability of winning the raffle is 0.3.
a Give a reason why Megan may be right.
b Give a reason why Megan may be wrong.

14 A ball is taken at random from a bag containing 12 balls, of which b are black.
a Write down, in terms of b, the probability that the ball will be:
  i black
  ii not black.
b When a further 6 black balls are added to the bag, the probability of getting a black ball is doubled. Work out the value of b.

15 A fair tetrahedral dice (4-sided, numbered 1 to 4) and an ordinary dice are each rolled. A win occurs when the number on the ordinary dice is greater than or equal to the number on the tetrahedral dice. Find the probability of a win.

16 There are two parts to a driving test: the theory test and the practical test.
You must pass the theory test before you pass the practical test.
Salma plans to take her driving test.
The probability that Salma will pass her theory test is 0.85.
The probability that she will pass her practical test is 0.65.
Work out the probability that Salma will pass her driving test.

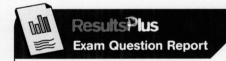

ResultsPlus
Exam Question Report

90% of students answered this sort of question poorly.

17 A naturalist wants to find out an estimate of the number of adult rabbits there are in a rabbit warren.
He catches 24 of the adult rabbits and marks them with some dye.
He returns the rabbits to the warren.
A few days later he catches 24 adult rabbits from the warren and finds that 6 are marked with the dye.
a Work out an estimate of the number of rabbits in the warren.
b Suggest the sort of assumptions you needed to make when you worked out this estimate.

18 A fruit machine has three independent reels and pays out a jackpot of £1000 when three raspberries are obtained. Each reel has 12 pictures of fruit. The first reel has four pictures of raspberries; the second reel has three pictures of raspberries and the third reel has five pictures of raspberries.
Find the probability of winning the jackpot.

A

19 Sarah and Jim each have a number of flower bulbs. Sarah has 4 daffodil bulbs and 5 hyacinth bulbs, and Jim has 3 daffodil bulbs and 7 hyacinth bulbs. They each pick one of their bulbs at random.
a Draw a tree diagram to show all the possible outcomes.
b Use your tree diagram to work out the probability that neither of the bulbs is a hyacinth.

A03

20 A fair game is one in which everyone has the same probability of winning.
For example, if you toss a fair coin, you have an equal chance of getting heads or tails.
Consider the following games. Are they fair? You should perhaps play them to understand how the game works. You must explain your answer.

a Three horses run a race over ten lengths. You toss two coins to see which horse moves.
- Horse A moves 1 length if you toss 2 heads.
- Horse B moves 1 length if you toss 2 tails.
- Horse C moves 1 length if you toss a head and a tail.
The horse that completes ten lengths first is the winner.

b Twelve horses run a race over ten lengths. You roll two dice to see which horse moves. For example:
- if you throw a 4 and a 3, Horse 7 moves 1 length
- if you throw a 6 and a 4, Horse 10 moves 1 length.
The horse that reaches ten lengths first is the winner.

A03

21 To play a lottery game, you choose six different numbers between 1 and 40.
Show that the probability of choosing all six numbers correctly is about 1 in 4 million.

A02
A03

22 A local garage donated a prize to Probability Junior School for its summer fair, thinking that it would be impossible to win.
To win the prize, someone had to roll six 6s from six dice. The entry fee was 10p, with money going to the school fund. If no one won then the garage kept the car.
Calculate the chances of the car being won.

A*A02
A03

23 The names Justin, Kayla, Hasan, Jessica, Amanda and Dave are each written on a piece of paper and placed in a hat. Two names are taken at random from the hat.
Work out the probability that the names are both boys' names.

A03

24 A bag contains 3 red sweets, 2 green sweets and 4 yellow sweets. Two sweets are taken at random from the bag. Work out the probability that:
a both sweets will be red
b both sweets will be the same colour
c the sweets will be different colours.

Results**Plus**
Exam Question Report

72% of students answered this sort of question poorly because they assumed the sweets were replaced, or they did not consider every scenario.

25 A fruit machine has three reels. Each reel has 10 pictures of fruit.
The table below gives information about the numbers of pictures of apples, pears, cherries and lemons on each of the reels.

	Apple	Pear	Cherry	Lemon
Reel 1	2	3	4	1
Reel 2	2	2	3	3
Reel 3	1	3	3	3

The fruit machine can be programmed to give a prize when particular fruit show on the reels, e.g. when three cherries show on the reels the fruit machine pays 10 tokens.
Programme the fruit machine to give prizes. Decide on the different combinations of fruit that will get a prize, the amount of the prize, and how many tokens a player will need to pay to play the fruit machine.

INTERPRETING AND DISPLAYING DATA

This question helps students develop their ability to choose and apply a method (AO2) and their ability to find an estimate by interpreting a graph (AO3). The AO2 skills are developed throughout the question by deciding what type of sampling to use, the choice of diagram to display the data and in the method of estimating between class intervals. The AO3 skills occur when estimating the number of boys who were between 152 cm and 156 cm tall from a cumulative frequency graph or a frequency density graph.

Example

Alan is doing a survey of the heights of 50 boys from his school.

He wants his sample to be fair and unbiased.

a Suggest a method he can use to take his sample.

b Explain how this method avoids bias.

The table shows information from his survey.

c Use a suitable diagram to display this data.

d Estimate how many boys were between 152 cm and 156 cm tall.

Height of boys in cm	Frequency
$140 \leqslant h < 145$	8
$145 \leqslant h < 148$	11
$148 \leqslant h < 150$	20
$150 \leqslant h < 154$	9
$154 \leqslant h < 160$	2

Solution

a Random stratified sample taking the year groups as strata within the sample.

b This method avoids bias by ensuring the number of boys selected from each year group is proportional to the size of the group. This ensures that one year group is not over- or under-represented.

c

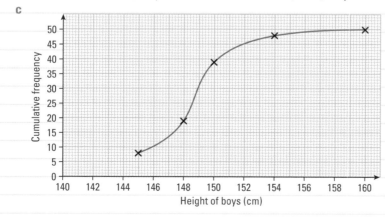

d 4 boys

Now try these

1   A geologist has taken a sample of pebbles from an area of interest.
    The table below shows some information about the weights of the pebbles.

Weight (w grams)	Frequency (f)
$0 \leqslant w < 20$	1
$20 \leqslant w < 30$	14
$30 \leqslant w < 40$	21
$40 \leqslant w < 50$	29
$50 \leqslant w < 60$	19
$60 \leqslant w < 70$	10
$70 \leqslant w < 80$	6

Estimate the number of pebbles that weigh between 54 g and 62 g.

2   The cumulative frequency curve shows the time
    taken for some workers to complete a task.
    Estimate the number of workers who took
    20–30 minutes to complete the task.

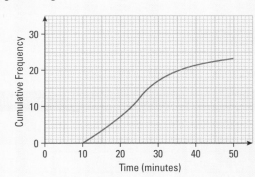

3   Within a radius of 5 miles from Sue's home there are 5 state secondary schools.
    Avon has 850 students, Moorside has 986 students, Heaton has 1296 students, Moortop has 1138
    students and Brambell has 1450 students.
    Explain how a sample of 60 students could be taken to give a fair representation of all of the schools.

4   The histogram below shows the waiting time for patients to
    be seen one morning by a doctor in a health centre.

    Exactly 20 people waited between 10 and 15 minutes
    for an appointment.
    No one waited more than 40 minutes.
    How many patients were seen by the doctor that morning?

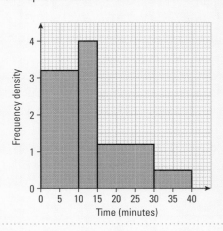

AVERAGE & RANGE

The following question helps you to develop both your ability to select and apply a method (AO2) and your ability to solve problems using your skills of interpretation and proof (AO3). The AO2 skills are developed as there is more than one way of working out the number of average size people that should be allowed in the lift. You could choose to work with the mean and range or the median and interquartile range. Your AO3 skills are needed in this question as you will need to give a reasoned explanation for your answer.

Example

One hundred people, selected at random, were weighed.
The results were put in a frequency table.

The maximum weight limit for the lift is 700 kg.
How many people of average size should be allowed in the lift?
Give a reason for your answer.

Weight (w kg)	Frequency (f)
$50 \leqslant w < 60$	6
$60 \leqslant w < 70$	29
$70 \leqslant w < 80$	45
$80 \leqslant w < 90$	19
$90 \leqslant w < 110$	1
Total	**100**

Solution

Student could choose to use

Weight (w kg)	Frequency (f)	mid-value	mid-value $\times f$	Cumulative frequency
$50 \leqslant w < 60$	6	55	330	6
$60 \leqslant w < 70$	29	65	1885	35
$70 \leqslant w < 80$	45	75	3375	90
$80 \leqslant w < 90$	19	85	1615	99
$90 \leqslant w < 110$	1	95	95	100
Total	**100**		**7300**	

Mean – best answer for safety reasons as it uses all of the values and allows for any extremes.
Mean is $\frac{7300}{100} = 73$ kg $\frac{700}{73} = 9.59$
Allow 9 people to use the lift with a 0.59 kg safety margin.

Mode – this is the most frequently occurring group.
Modal group is $70 \leqslant w < 80$ using the upper limit $\frac{700}{80} = 8.75$
Allow 8 people to use the lift with a 0.75 kg safety margin.

Median – this is the middle person and occurs in the $70 \leqslant w < 80$ group.
Estimate about 73 kg $\frac{700}{73} = 9.59$
Allowing 9 people in the lift with a 0.59 kg safety margin.

1 A is the set of data 1, 2, 4, 5, 8, 10.
B is the set of data 3, 4, 6, 7, 10, 12.
C is the set of data 2, 4, 8, 10, 16, 20.
 a Compare the mean of A and the mean of B.
 b Compare the range of A and the range of B.
 c Compare the mean of A and the mean of C.
 d Compare the range of A and the range of C.
 e Write down a data set which has the same mean as B but twice the range.

2 The table below shows the marks given in two Maths tests.

| Boys | 62 | 75 | 67 | 81 | 79 | 91 | 69 | 73 | 85 | 81 |
| Girls | 74 | 69 | 83 | 83 | 78 | 68 | 88 | 81 | 68 | |

Compare the distributions for boys and girls.
Farida was absent. What is the minimum mark she needs to score when she takes the test to keep the girls average better than that of the boys?

3 The mean mark for 10 pupils in an English Language Examination was x.
5 of the students were awarded an extra 4 marks for the quality of their written communication.
What difference does this make to the average for the group?

4 The table below shows the Maths and English marks for a group of 50 pupils.

Mark	21 – 30	31 – 40	41 – 50	51 – 60	61 – 70	71 – 80	81 – 90	91 – 100
Maths	1	2	8	11	14	8	4	2
English	1	1	3	20	14	9	1	1

Choose a suitable diagram to display these results.
Compare the distribution of the two sets of marks.

5 Kay noted the number of hours of sunshine in May and June.
The results of her survey are given below.

Hours of sunshine per day	May	June
Mean (hrs)	9.6	8.4
Range	12	16

 a Compare the number of hours of sunshine for May and June.
 b Kay chose to use the mean and range instead of using the median and interquartile range.
  i Give one reason why this might be a good choice.
  ii Give one reason why it might be better to use the median and interquartile range.

The following question helps you develop both your ability to select and apply a method (AO2) and your ability to analyse and solve problems (AO3). The AO2 skills are used in the first part of the question. Setting up an equation is the most efficient way to work out the probability of choosing a mint but some students may choose to estimate with trial and improvement. Your AO3 skills will be developed as you will have to work out how to solve the problem.

Example

A bag of sweets contains mints, toffees and creams.
The probability of choosing a toffee is $\frac{1}{3}$.
The probability of choosing a cream is x.
The probability of choosing a mint is $2x$.
What is the probability of choosing a mint?
Give your answer as a numerical value.

Solution 1

One strategy is to set up an equation.

$\frac{1}{3} + x + 2x = 1$

$3x = \frac{2}{3}$

$x = \frac{2}{9}$

Probability of choosing a mint is

$2x = \frac{4}{9}$

> This question requires you to work with algebra and probability

Solution 2

Alternatively we can estimate the number of sweets. As thirds are involved choose a multiple of 3. Select a total number of sweets which is a multiple of 3 e.g. 36

toffees = 12 sweets
mints + creams = 24 sweets.

P (mint) = 2P(cream) ⟵ We can compare quantities.
So 16 mints to 8 creams

So P(mint) = $\frac{16}{36} = \frac{4}{9}$

1 In a box of chocolates, there are 3 types of chocolate.
The probability of a cream is $2x$.
The probability of a toffee is $3x$.
The probability of a hard centre is $4x$. Calculate the probability of choosing a toffee.

A02 C

2 A bag contains 20 sweets. x of the sweets are chocolates, the rest are toffees.
Mona takes a toffee from the bag and eats it.
She then offers the bag to Sam who eats a sweet.
Explain why the probability of Sam eating a chocolate is not $\frac{x}{20}$.

A03

ResultsPlus
Hint

How many sweets are left in the bag when Mona has eaten one?

3 A drawer contains a number of black and grey socks.
The probability that the first sock Ali pulls from the drawer will be black is x.
Explain why the probability of pulling a second black sock from the drawer is not x.

A03

4 A sandwich shop has 3 types of sandwiches; ham, cheese and prawn.
The probability of a customer choosing ham is $\frac{1}{2}$.
The probability of a customer choosing cheese is x.
The probability of a customer choosing prawn is $3x$.
Mark chooses a sandwich.
What is the probability of him choosing prawn?
Give your answer as a numerical value.

A02 A
A03

5 The probability that a train arrives on time is 70%.
The probability that it arrives early is x.
The probability that it arrives late is $2x$.
Calculate the numerical value of it arriving late.

A02 A03

ResultsPlus
Hint

Probabilities can be fractions, decimals or percentages.

6 The top floor of a block of flats can be reached by two lifts.
The probability of both of the lifts working is 0.85.
The probability of only one lift working is $2x$.
The probability of none of the lifts working is x.
Calculate the numerical probability of only one lift working.

A02

7 In an athletics competition Dan enters for two events.
The probability of Dan winning one event and losing one is $2x$.
The probability of him winning two events is x.
The probability of him not winning any events is $5x$.
Calculate the probability that he does not win any event.

A02

FS

Snowdon is the highest mountain in Wales. To get to the summit of Snowdon you can walk or go by train. Read the information below then answer the questions opposite.

Here are 5 possible routes up Snowdon.

Route	Distance (km)	Ascent (m)	Average time taken (hrs)	Level of difficulty
Llwybyr Llanberis Path	9	1020	3.5	
Snowdon Ranger	7.5	1100	3	
Miner's Track	7	930	2.5	
South Ridge	7	890	2.5	
Lliwedd	6	970	4	

Fact 1: 1 foot = 0.3048 metres.

Fact 2: The height of Snowdon in feet is 3560 ft.

Here are the train fares.

TRAVEL FARES

	Llanberis–Clogwyn Return fare	Llanberis–Clogwyn Single fare	Llanberis–Summit Return fare	Llanberis–Summit Single fare
Adult	£17	£12	£22	£16
Senior / Student	£12	£8	£17	£13
Child	£14	£9	£15	£10
Early bird Adult*	£9	£7	£13	£12
Early bird Child*	£6	£5	£6.50	£5.50

*The Early Bird train departs at 9 am.

1. Jon, Sarah and their two children, Poppy and Mark, are going to buy single fare tickets from Llanberis to the summit of Snowdon. They can go by the 9 am train or the 11.30 am train. Which train should they catch to get the cheapest fares and by how much?

2. Rashid and Chelsea each walk to the summit of Snowdon. Rashid takes the South Ridge route and starts at 09:45. Chelsea takes the Llwybyr Llanberis Path route and starts at 09:30. They both arrive at the summit at 12:30. Rashid says that he is the faster walker. Is he right? Give a reason for your answer.

3. a Draw a chart or graph to convert feet into metres.
   b The highest mountain in Scotland is Ben Nevis. The height of Ben Nevis is 1344 m. Which is higher, Ben Nevis or Snowdon?

Exercise is not based on real data.

LINKS

⊙ You need to be able to read data from tables in **Question 1**. You learnt this skill in **Chapter 1**.

⊙ You need to work out time intervals for **Question 2**. You learnt this in **Chapter 1**.

⊙ **Question 3** asks you to draw a conversion graph. You learnt this in **Chapter 4**.

FS MUSIC SALES

L istening to music is one of the most popular pastimes for people of all ages. In the twenty-first century music can be purchased in a number of formats. Read the information below, then answer the questions.

QUESTION

1. Here is some information about music sales between 2006 and 2009 in the UK. Explore the trend in sales for CDs and downloads between 2006 and 2009. You must show your working.

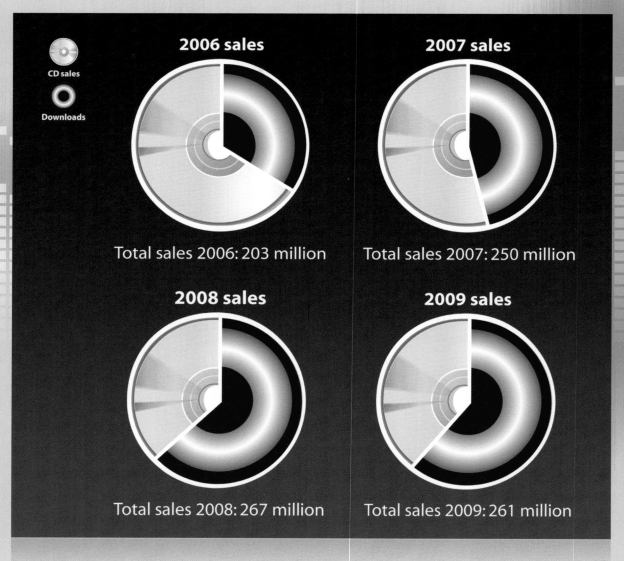

CD sales
Downloads

2006 sales

Total sales 2006: 203 million

2007 sales

Total sales 2007: 250 million

2008 sales

Total sales 2008: 267 million

2009 sales

Total sales 2009: 261 million

2. The tables show a breakdown by age for the sales of CDs and downloads in 2008..

Simon wants to show this information in a suitable diagram or graph that will allow people to compare the information visually. Produce an example of how Simon could do this.

CD Sales

Age (years)	CDs (millions)
$0 \leqslant x < 15$	13
$15 \leqslant x < 30$	22
$30 \leqslant x < 45$	37
$45 \leqslant x < 60$	13
$60 \leqslant x < 75$	8
$75 \leqslant x < 90$	5

Downloads

Age (years)	Downloads (millions)
$0 \leqslant x < 15$	42
$15 \leqslant x < 30$	63
$30 \leqslant x < 45$	29
$45 \leqslant x < 60$	19
$60 \leqslant x < 75$	10
$75 \leqslant x < 90$	6

QUESTION

3. Prepare a short questionnaire to obtain information about how much people spend on music and what format they buy. You need information about their age and gender as part of the information you collect.

Design a suitable data collection sheet for collating the information.

LINKS

⦿ For **Question 1** you need to be able to interpret pie charts. You learnt how to do this in **Chapter 3**.

⦿ For **Question 2** you need to find the best way of visually representing data. You learnt about the various ways of doing this in **Chapter 3**.

⦿ You learnt how to design questionnaires in **Chapter 1**. You will need to do this in **Question 3**.

Exercise is not based on real data.

Answers

Chapter 1

1.1 Get Ready

a Ask each classmate how much lunch money they get and calculate the mean of this data.

b Use an internet air ticketing site or the Manchester airport site.

c Search the internet for a government site that gives information on voting figures.

Exercise 1A

1. **a** secondary **b** secondary **c** primary
2. **a** quantitative **b** qualitative **c** quantitative
   d quantitative
3. **a** discrete **b** continuous **c** continuous
   d discrete
4. 26 metres
5. 0.92 metres
6. 280
7. 96. He will get 45 pieces from the first length and 51 from the other. He will have pieces 6 metres and 8 metres left. Although this is more than 10 metres he can not use this flex as it will not form a piece 10 metres long without a join.
8. **a** 469 centimetres
   b 4.69 metres
   c 800 − 469 is 331 cm. This is enough to make two extra skirts − one for Latika plus one for either Nirupa or Saria.
9. **a** Drug A is effective at curing malaria OR Drug A is not effective at curing malaria.
   b Collect data on patients that have been treated with Drug A and those that haven't.
10. No she has not enough flour. She needs $125 \times 15 = 1875$ g and she only has 1500 g.
11. **a** 25 glasses **b** 3 bottles **c** £6.60

1.2 Get Ready

If the answers are written down: 15 seconds
If the answers are given orally (i.e. one student at a time): number of students in class $\times$ 15 seconds.

Exercise 1B

1. For example, rolling a dice or using the random function on a calculator or from a random number table
2. A fraction of the population is chosen at random.
3. For example, assign the numbers 1 to 60 to the workers, then take the first eight different numbers under 61 that are generated by the calculator
   (21, 32, 54, 34, 26, 45, 35, 22)

1.3 Get Ready

1. **a** $\frac{15}{25} = \frac{3}{5}$ **b** $\frac{10}{25} = \frac{2}{5}$
2. **a** 80 **b** $\frac{3}{8}$

Exercise 1C

1. 23 boys, 27 girls
2. Randomly select 15 employees with less than six months' experience and 40 employees with more than six months' experience.
3.

	Office workers	Factory floor workers	Managers
Females	5	25	1
Males	8	49	2

They should be picked by simple random sampling.

1.4 Get Ready

1. **a** 4 **b** 6
2. 卌 |

Exercise 1D

1. **a**

Vehicle	Frequency
Car	卌 卌 卌 卌 ||
Bus	卌
HGV	||||
Bike	||||
Motorbike		

2. 6, 1, 11, 9, 18
3. 12.6, 1.0, 2.5, 1.5, 8.4
4. **a** Train B **b** Train A
   c Amelia is wrong because she has read 14:30 as 4.30 pm.
5.

Number of DVDs bought	Frequency
0–3	9
4–7	11
8–11	6
12–15	4

b Motorbike **c** Car

6. **a**

Weight	Frequency
$57 \leqslant w < 60$	7
$60 \leqslant w < 63$	3
$63 \leqslant w < 66$	5
$66 \leqslant w < 69$	5
$w \geqslant 69$	4

b $57 \leqslant w < 60$ **c** $60 \leqslant w < 63$

1.5 Get Ready

Tallying

Exercise 1E

1 It is a biased question.

2 A: open, B: closed, C: open, D: closed

3 a No option for dissatisfied customers.
New suitable question:
What do you think of the new amusements?
Very good ☐ Good ☐ Satisfactory ☐ Poor ☐

b Options overlap.
New suitable question:
How much money would you normally expect to pay for each amusement?
£5–£7 ☐ £7.01–£8 ☐ more than £8 ☐

c Not clear what the options mean.
New suitable question:
How often do you visit the park each year?
0–2 times ☐ 3–5 times ☐ 6–8 times ☐
more than 8 times ☐

4 Do you like the new layout? Yes/No

1.6 Get Ready

1 a 15 girls **b** 6 students

Exercise 1F

1 a

	Plain	Salt and Vinegar	Cheese and Onion	Total
Males	7	7	14	28
Females	5	6	12	23
Total	12	13	26	51

b 13 **c** 51

2 a

	Orange Juice	Grapefruit Juice	Total
Men	22	8	30
Women	18	12	30
Total	40	20	60

b 40

3 a

	Supervisors	Office staff	Shop floor workers	Total
Males	10	3	82	95
Females	2	11	38	51
Total	12	14	120	146

b 51 **c** 146

Exercise 1G

1 a £385 + £385 + £300 = £1070
b Either by going in Jan 2010 or by going for a shorter time.

Exercise 1H

1 Yes, only one area sampled

2 A: Biased. Not everyone in the hospital's area has a chance of being asked.
B: Biased. Only people with phones have a chance of being asked and only in 10 towns, the sample is too small.
C: Not biased.
D: Biased. Only people already using the recycling facility are being asked.

3 Pupil's own discussions

1.8 Get Ready

The internet, supermarkets, high street shops

Exercise 1I

1 a 41 000 tonnes **b** Cars **c** 2004
d Cars

2 a 6.3 days **b** May **c** May **d** May

3 a 799 **b** Stays the same
c More dairy **d** Numbers have decreased

Review exercise

1 a 7 cm, 8 cm, 6 cm, 6 cm, 7 cm
b 4.6 cm, 4.0 cm, 4.6 cm, 5.7 cm, 4.0 cm

2 a It does not allow for sending no text messages. It does not include a time frame, e.g. per week.
b It only includes people of one age.

3 a The categories overlap. It does not include a time frame, e.g. per day. It does not allow for people who use their computer for more than 6 hours.
b It only includes people of one age.

4 a

	Time
Courtney leaves home	0648
Train departs Oxenholme	0708
Train arrives Glasgow	0914
Courtney arrives for interview	0949
Interview finished	1115
Train leaves Glasgow	1600
Train arrives Oxenholme	1742
Courtney arrives home	1800

b The train does not stop at Penrith. It is quicker travelling between Carlisle and Glasgow.

5 a £3948
b If he wants to go business class he could reduce the time away to 3 days. He could save a lot of money by going Economy Class. He could change hotel and go to the Metro Hotel.

6 a It does not allow for never visiting the cinema. It is hard to decide what the categories mean. It does not include a time frame, e.g. per month.
b On average, how many times do you go to the cinema each month?
0–1 times ☐ 2–3 times ☐ 4–5 times ☐
more than 5 times ☐

7 On average, how often do you shop at this supermarket each month?

0 – 1 times ☐   2 – 3 times ☐   4 – 5 times ☐
more than 5 times ☐

8 On average, how many emails do you send each week?

0 – 5 ☐   6 – 10 ☐   11 – 15 ☐   16 – 20 ☐
more than 20 ☐

9 It only includes women. It does not include people who never go to the cinema.

10

Animal	Tally	Frequency
Lions		
Tigers		
Elephants		
Monkeys		
Giraffes		

11

Country	Tally	Frequency
France	ℍ	5
Spain	ℍ II	7
England	IIII	4
Italy	IIII	4

12 **a** The first question does not allow for people who never visit the park and it is hard to decide what each category means.
The second question has overlapping categories.

  b On average, how often do you go to the County Park each month?

  Never ☐   1 – 3 times ☐   4 – 6 times ☐
  more than 6 times ☐

  How old are you?

  0 – 10 years ☐   11 – 20 years ☐   Over 20 years ☐

13 **a** Biased, because it only includes those working on the night shift.

  b Not biased, because it uses a simple random sample.

  c Biased because the question starts with 'Do you agree…'

14 6 girls

15 12 boys

16 **a** 5 students   **b** 26 students

17 19 students

18

	London	York	Total
Boys	23	14	37
Girls	19	24	43
Total	42	38	80

Chapter 2

2.1 Get Ready

a 72   **b** 65

Exercise 2A

1 **a** £1179.51   **b** 204   **c** 40

2. 12 cm

3 **a** 81   **b** 125   **c** 13
d 7   **e** 17   **f** 27

4 The walls for one room are $7\,m^2$, $7m^2$, $8\,m^2$, $8\,m^2$. This means each room is $30\,m^2$. Three rooms are $90\,m^2$. Ceilings are each $14\,m^2$. So three is $42\,m^2$. This is a total of $132\,m^2$ to paint. The cost will be £112.20. The labour is £330. The total cost is £442.20.
Max has enough money.

5 $8.50 - 2.65 = 5.85$. After making one owl box he has 5.85 m left.
He could make two more owl boxes leaving him with 55 mm left.
He could make 4 robin boxes leaving 5 mm spare
He could make one more owl box and two robin boxes leaving 30 mm spare.
The least waste is to make the one owl box and 4 robin boxes.

2.2 Get Ready

a 1 2 3 4 5 5 7 8 8 9 10 12   **b** 3.5, 3.5, 4.5, 4.6, 6.2, 8.7, 12.5

Exercise 2B

1 2
2 16 litres
3 **a** 16   **b** 16
4 227

2.3 Get Ready

a 5   **b** $5a$

Exercise 2C

1 $6x$
2 **a** 4   **b** $4 + (p - 1)$ or $3 + p$
3 $x + y$
4 ps pence
5 **a** expression **b** equation **c** formula **d** identity
6 348
7 **a** $3x + 7$ **b** $3y + £6.00 = 6y + £3.00$ so $y = £1$

2.4 Get Ready

1 **a** 44   **b** 20.4

Exercise 2D

1 3
2 **a** 136   **b** 134
3 £1292
4 **a** 10   **b** 20   **c** 24

2.5 Get Ready

a Median $= 20$
Mode $= 20$
b Mean $= 21$ to 2 s.f.

Exercise 2E

1 a Mean £47, mode £17, median £22
  b The median is best. The mean has been affected by one high value and the mode is the lowest value.
2 One advantage from: Is the most popular measure. Can be used for further calculations. Uses all the data. One disadvantage from: Affected by extreme values. Actual value may not exist.
3 a 28
  b 38
  c 64 (to the nearest whole person)
  d The mean because it is the highest average.

2.6 Get Ready

a 81   **b** 6

Exercise 2F

1 a £82.80   **b** £13.80
2 a 21.16   **b** 704.969   **c** 45.6
  d 58   **e** 20   **f** +96
3 a 8   **b** 8.267

2.7 Get Ready

Number	1	2	3	4	5	6
Frequency	1	4	4	3	3	1

Exercise 2G

1 a Frequency $\times$ number of siblings: 0, 8, 18, 12, 12, 0, 12, 7
  Total $f = 30$
  Total $f \times x = 69$
  b 2   **c** 2   **d** 2.3
2 a 104   **b** 104   **c** 104
3 a 17   **b** 18   **c** 18.5

2.8 Get Ready

Class interval	Frequency
1–4	11
5–9	5
10–14	7
15–19	7

Exercise 2H

1 a £281−£320   **b** £321−£360
2 a $0.45 \leqslant x < 0.50$ **b** $0.40 \leqslant x < 0.45$
3 a $70.0 \leqslant x < 70.1$ **b** $69.9 \leqslant x < 70.0$

2.9 Get Ready

a 60   **b** 0.8   **c** 0.000 55

Exercise 2I

1 42.5   **2** 181.0 seconds   **3** 54.23 seconds

2.10 Get Ready

1 14 16 16 18 21 23 27 32 38 43 45 49
2 36.0 kg 43.4 kg 43.5 kg 49.9 kg 56.2 kg 56.2 kg

Exercise 2J

1 Median
2 a $Q_1 = 50$, $Q_2 = 62$, $Q_3 = 70$ **b** 20   **c** 42
3 a $Q_1 = 8$, $Q_2 = 13$, $Q_3 = 19$ **b** 11   **c** 19
4 a $Q_1 = 32$, $Q_2 = 45$, $Q_3 = 52$ **b** 20   **c** 47

Review exercise

1 a 4.5   **b** 3.6   **c** 4
2 a 20 biscuits   **b** 25 g
3 a Mode = £4, median = £5, mean = £9
  b The median, because the mode is close to the lowest value and the mean is affected by the single large amount of £38.
4 a One advantage from: unaffected by extreme values; can be used with qualitative data.
    One disadvantage from: may be more than one mode; may not be a mode.
  b Advantage: not influenced by extreme values.
    Disadvantage: actual value may not exist.
  c One advantage from: can be used for further calculations; uses all the data.
    Disadvantage: affected by extreme values.
5 a 24   **b** 3.6
6 a The mode is 12, the number of rooms that occurs most frequently. Ali has given the maximum number of rooms.
  b 6.3 to 1 d.p.
7 2.4 to 1 d.p.
8 45
9 7.7
10 a 10.5–10.7   **b** 10.2–10.4
11 a

Class Interval	Frequency (f)	Class mid-point	$f \times x$
$26 \leqslant w < 29$	4	27.5	110
$29 \leqslant w < 32$	7	30.5	213.5
$32 \leqslant w < 35$	15	33.5	502.5
$35 \leqslant w < 38$	12	36.5	438
$38 \leqslant w < 41$	2	39.5	79
Totals	40		1343

  b 33.6 kg to 3 s.f.
12 84.8 mm to 3 s.f.
13 19 minutes
14 a $30 < t \leqslant 40$ **b** 27.3 minutes
15 13.0 minutes to 3 s.f.
16 Year 11 - 26 pets (Year 9 – 76 pets, Year 10 – 72 pets)
17 a $Q_1 = 42$ kg, $Q_2 = 47$ kg, $Q_3 = 49$ kg
  b 7 kg   **c** 11 kg
18 a 18 minutes   **b** 15 minutes
  c Before: $Q_1 = 14$ minutes, $Q_3 = 22$ minutes
    After: $Q_1 = 11$ minutes, $Q_3 = 19$ minutes
  d Before: interquartile range = 8 minutes
    After: interquartile range = 8 minutes

e Before the introduction of the traffic management scheme the mean time taken to travel to work was higher than after the scheme was introduced. The interquartile range stayed the same, so the spread of times was similar before and after the scheme started.

19 Mean $= \dfrac{300\,000}{10} = £30\,000$. The owner could say that the average salary is £30 000 and it is high enough.
Mode/median $= £10\,000$. The workers could say that the average salary is £10 000 and too low.

20 If 1 dog in 100 had three legs, then the mean number of legs $= \dfrac{399}{100} = 3.99$
The majority of dogs have four legs.

Chapter 3

3.1 Get Ready

1 360
2 90
3 a 120 **b** 45 **c** 60

Exercise 3A

1 60°
2 a i 60⁰ **ii** 10 minutes **b** 2.5 hours
3 160⁰

Exercise 3B

1
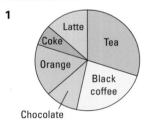

2
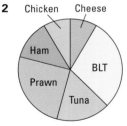

3
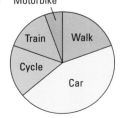

Exercise 3C

1 a Golf **b** Athletics **c** 45 **d** 30
2

3.3 Get Ready

a 50, 54, 65, 72
b 4.0, 4.3, 4.4, 4.6
c 0.01, 0.1, 0.11, 0.12

Exercise 3D

1 a 36 **b** 28 **c** 47 **d** 18, 36 **e** 18
2 a

0	4	5	6	7	7	8	9	9	9	9
1	0	2	4	5	7					
2	1	4	8							
3	0									

Key 2 | 1 stands for 21
b 9 minutes **c** 9 minutes **d** 26 minutes
e 7 minutes, 17 minutes **f** 10 minutes
3 a

5	2	6	9						
6	3	3	4	5	5	8	8	8	9
7	2	4	4	4	4				
8	2	3	5	8					
9	2	4							

Key 7 | 2 stands for 72
b 74 km **c** 69 km **d** 42 km
e 64 km, 82 km **f** 18 km

3.4 Get Ready

a B occurs more frequently than A, which occurs more frequently than C.
b The red category accounts for a higher proportion of the total than the green or blue categories. Red: 50%, green and blue: 25% each.

Exercise 3E

1 a Food **b** Pension **c** 25%
2 a 70 g **b** 10 g **c** Fruitbix
3 a Saturday **b** Thursday **c** 40

3.5 Get Ready

a 3 **b** 16 **c** 25

Exercise 3F

1

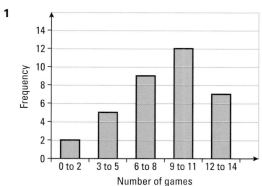

2 a $135 \leqslant w < 140$ **b** $140 \leqslant w < 145$

c

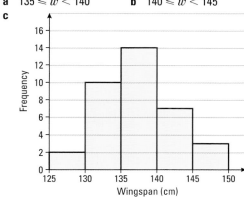

3 a $139 \leqslant w < 141$ **b** $137 \leqslant w < 139$

c

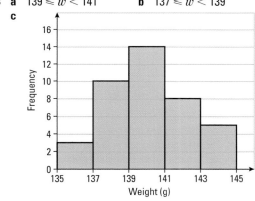

3.6 Get Ready

a 5   **b** 17.5   **c** 115.5

Exercise 3G

1

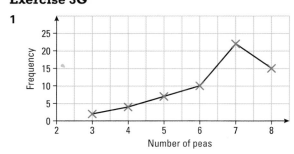

2 a $70 \leqslant d < 80$

b, c

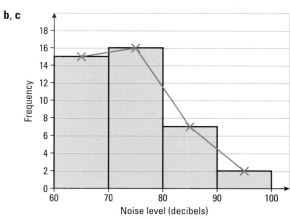

3

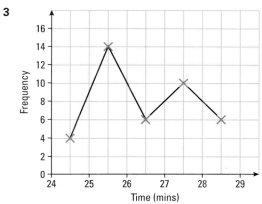

4 Girls, because their mode is 5 minutes and the boys' mode is 9 minutes.

Exercise 3H

1 a

Lifetime (l hours)	Frequency	Class width	Frequency density
$10 \leqslant l < 15$	4	5	0.8
$15 \leqslant l < 20$	10	5	2
$20 \leqslant l < 25$	20	5	4
$25 \leqslant l < 30$	15	5	3
$30 \leqslant l < 40$	6	10	0.6

b

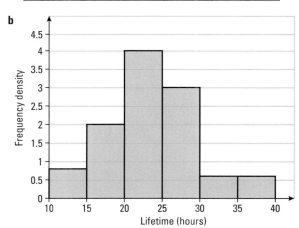

Answers

2

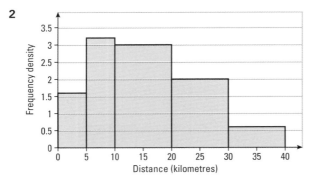

25

3 a

Age (y years)	Frequency
$0 < y \leqslant 5$	10
$5 < y \leqslant 10$	28
$10 < y \leqslant 20$	49
$20 < y \leqslant 40$	48
$40 < y \leqslant 70$	60

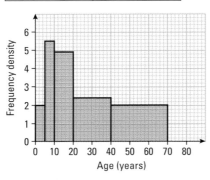

b 101

Exercise 3I

1 a Cumulative frequency: 3, 10, 20, 35, 43, 48, 50
b

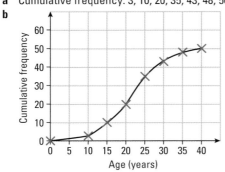

2 a 11 **b** 19 **c** 33
3 a i 22 **ii** 74
b 110
c 18%

3.9 Get Ready

a 9 and 12
b 18 and 20

Exercise 3J

1 a 42 **b** 33, 50 **c** 17 **d** 70
2 a £6000 **b** £4000, £7500 **c** £3500
3 a Median = £242 000, Q_1 = £222 000, Q_3 = £256 000
b Range = £120 000, interquartile range = £34 000

3.10 Get Ready

Median = 15
Lower quartile = 8
Upper quartile = 25
Interquartile = 17

Exercise 3K

1

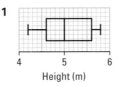

2

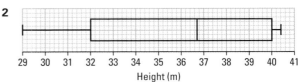

3 a Male: min = 10, Q_1 = 20, Q_2 = 40, Q_3 = 50, max = 80
Female: min = 10, Q_1 = 30, Q_2 = 50, Q_3 = 60, max = 80
b

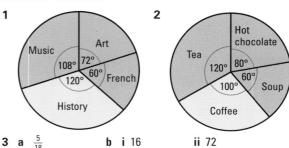

The female members are older on average (higher median). The interquartile range is the same for males and females, but the range is slightly greater for females so their ages vary slightly more.

Review exercise

1 **2**

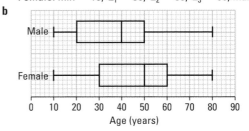

3 a $\frac{5}{18}$ **b i** 16 **ii** 72
c If many fewer students took the English exam then the number getting Grade D could be fewer than the number of students getting Grade D for mathematics, even though the proportion of the total is higher.
4 a 14 **b** 5

c

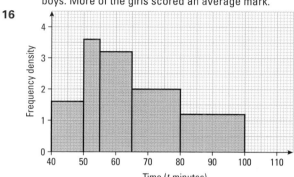

d Tuesday

5 a

4	6	8		
5	1	2	8	
6	0	3	6	8
7	4	7	8	9
8	7			

Key 5 | 1 means 51 kg

  b 5   **c** $87 - 46 = 41$ kg

6 30 mm

7

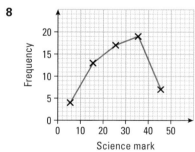

8

9 James. The angle for the combined proportion must be between the Year 9 and Year 10 angles.

10 On average, the prices are lower at Peter's garage than at John's garage, as the median is lower. The range and interquartile range are both smaller for Peter's garage so the spread is less than for John's garage. Both the cheapest and the most expensive cars come from John's garage.

11 a 88 people   **b** 38 years   **c** $54 - 22 = 32$ years

12 a

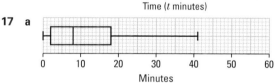

Girls' height (m)

b On average, the girls are taller than the boys, as the median is higher, but their heights are more variable as the range is larger. The interquartile ranges are the same so the variation in the middle of each group is similar.

13 From a cumulative frequency chart, the median for the men is £240. On average, the men spent more than the women.

14 a 23 kg is the value of Q_3. The heaviest bag weighs 29 kg.
   b 17 kg   **c** 13 kg   **d** 60 bags

15 a 21   **b** 10
   c Most of the high and low scores are scored by the boys. More of the girls scored an average mark.

16

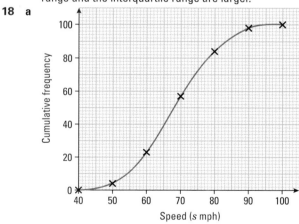

17 a

b The delays were greater on Saturday, as the median and the quartiles are all higher. There was also more variation in the delays on Saturday, since both the range and the interquartile range are larger.

18 a

b 88 mph   **c** $76 - 61 = 15$ mph

Chapter 4

Exercise 4A

1 a 7   **b** 5.5
2 a 96 cm²   **b** 4 cm
3 a

x	0	1	2	3	4	5
$y = 2x + 1$	1	3	5	7	9	11

Answers

b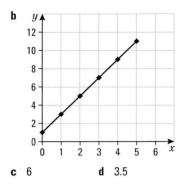

c 6 **d** 3.5

4.2 Get Ready

1 a $y = 3$ **b** $y = 5$ **c** $y = 8$

2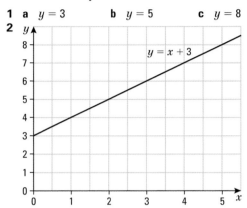

Exercise 4B

1 $\frac{1}{5}$ or 0.2

2 a By inspection **b** By inspection **c** By inspection

d 6

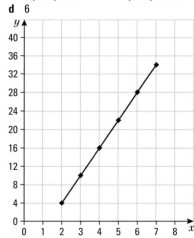

3 Gradient 2. This means that for every increase in body length of 2 cm the tail increases by 1 cm.

4 For every increase in litter size of 1 the mass of each piglet in the litter decreases by 50 g.

4.3 Get Ready

a 0.05 **b** 0.4

Exercise 4C

1 a 700 m
b They stopped twice; once for 1 minute then again for 2 minutes.
c 9.5 minutes
d 420 m

2 a Month 1; 95 m **b** 80 m
c Months 6 and 8. These correspond to June and August when it doesn't rain much.

3 a 18:00 hours, because people are making their evening meals
b 102 000 kilowatts
c 12:00, 16:36, 20:00
d 22:00

4.4 Get Ready

a A 63.5, B 64.25 **b** C 10.75 **c** D 54.9 **d** E 16.6

Exercise 4D

1 a

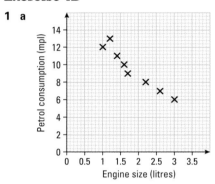

b Negative correlation
c The greater the engine size, the lower the petrol consumption.

2 a

b No correlation
c Screen size and selling price are not related.

3 a

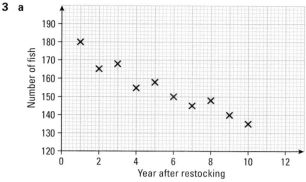

b Negative correlation
c The more years have passed since restocking, the smaller the number of fish.

4 a Related – the lighter the car, the faster it will go.
b Related – the longer the motorway, the greater the number of petrol stations.
c Unrelated
d Related – the greater the number of bicycles sold, the greater the number of cycle helmets sold.

Exercise 4E

1 a, c

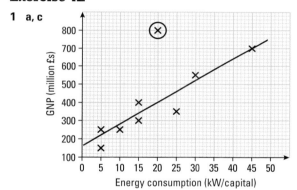

b (20, 800)

2 a, c

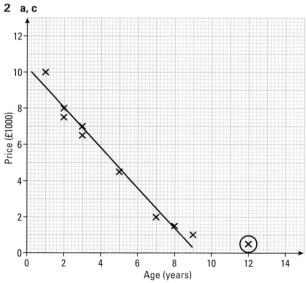

b Car prices cannot go negative, so the relationship only holds for cars up to 9 years old. Cars older than this will still be worth a small amount.

3 a, b, d

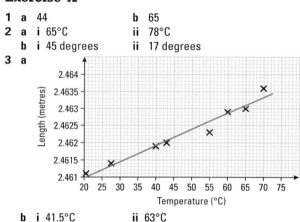

c Ottawa. The high and low temperatures are the wrong way round.

Exercise 4F

1 a 44   **b** 65
2 a i 65°C   **ii** 78°C
  b i 45 degrees   **ii** 17 degrees
3 a

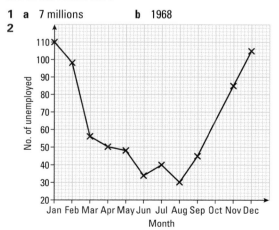

b i 41.5°C   **ii** 63°C
c 2.461 45 m

Review Exercise

1 a 7 millions   **b** 1968
2

b 65
c August, because more people go to the seaside in the summer holidays.

3 a, b

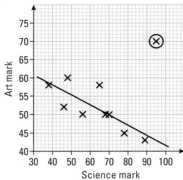

c Negative correlation

d The higher the science mark achieved by a student, the lower the art mark.

4 a, d

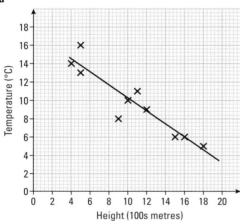

b Negative correlation

c The greater the height about sea level, the lower the temperature.

5 a, c

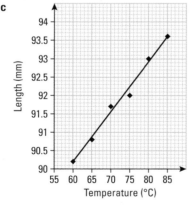

b Positive. As temperature rises so does the length.

d 0.14

e For every rise of 10 the length increase by approximately 0.14 mm

6 a The higher the price, the fewer the number of cameras sold.

b

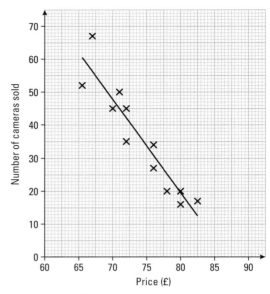

7 a As a car gets older, its value decreases.

b

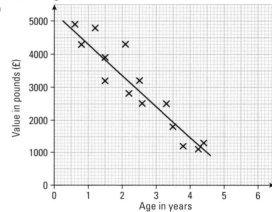

c £2450

d 1.9 years

8 a The larger the weight of a child, the greater its height.

b

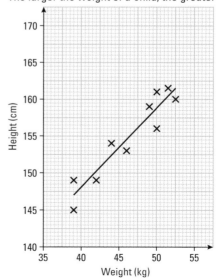

9 a, c

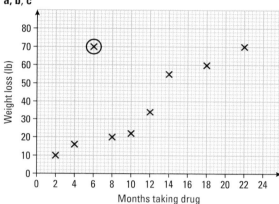

b Positive correlation

c The taller the athlete, the greater his or her weight. Using the line of best fit above, 74.2 kg

10 a, b, c

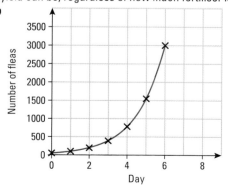

c This is unlikely to be a genuine piece of data as the weight loss is so high in only 6 months.

d Possitve correlation

e The drug is effective as the longer that the drug was taken, the greater the weight loss.

11 a Possitive correlation

b The greater the amount of fertiliser used, the higher the crop yield.

c 20 200 kg

d 2.5 kg per 80 m²

e No, because there will be a limit on how high the crop yield can be, regardless of how much fertiliser is used.

12 a, b

```
Number of fleas vs Day graph
3500
3000
2500
2000
1500
1000
500
0
  0   2   4   6   8
       Day
```

c $y = k^x$

13 a For every 100 m rise in height above sea level there is 1° drop in temperature.
OR For every 1 m rise in height above sea level there is 0.01° drop in temperature.

b At sea level the air temperature is 19°.

Chapter 5

5.1 Get Ready

1 0 **2** 1 **3** $\frac{1}{2}$

Exercise 5A

1 a 0.1, 0.6, 0.5, 0.46

b $\frac{2}{5}, \frac{3}{4}, \frac{9}{20}, \frac{18}{25}$

c 62%, 40%, 30%, 15%

2 $\frac{1}{10}$, 0.3, 48%, $\frac{3}{5}$

3 a $6\frac{1}{8}$ **b** 95

4 70%

Exercise 5B

1 a $\frac{2}{7}$ **b** $\frac{4}{7}$ **c** $\frac{1}{7}$

2 a $\frac{1}{6}$ **b** $\frac{1}{2}$ **c** $\frac{1}{13}$

d $\frac{1}{4}$ **e** $\frac{1}{26}$

3 a $\frac{5}{9}$ **b** $\frac{4}{9}$ **c** 0

4 a $\frac{1}{2}$ **b** $\frac{1}{2}$ **c** $\frac{1}{4}$

d $\frac{5}{8}$ **e** $\frac{3}{8}$

5 $\frac{1}{500}$

6 a $\frac{1}{3}$ **b** $\frac{4}{9}$ **c** $\frac{2}{9}$ **d** $\frac{2}{3}$

7 a $\frac{2}{11}$ **b** $\frac{1}{11}$ **c** $\frac{3}{11}$

d $\frac{4}{11}$ **e** 0

8 a

	Left handed	Right handed	Total
Boys	47	135	182
Girls	61	119	180
Total	108	254	362

b i $\frac{91}{181}$ **ii** $\frac{54}{181}$ **iii** $\frac{119}{181}$

9 a $\frac{1}{6}$ **b** $\frac{7}{12}$

10 0.3

Exercise 5C

1 a i $\frac{1}{36}$ **ii** $\frac{1}{9}$ **iii** $\frac{1}{6}$ **b** $\frac{1}{6}$ **c** $\frac{1}{9}$

2 a

		1	2	3	4	5	6
Coin	H	(1,H)	(2,H)	(3,H)	(4,H)	(5,H)	(6,H)
	T	(1,T)	(2,T)	(3,T)	(4,T)	(5,T)	(6,T)
		1	2	3	4	5	6

Dice

b i $\frac{1}{12}$ **ii** $\frac{1}{2}$ **iii** $\frac{1}{6}$

Answers

3 a

		Spinner A			
		1	**2**	**3**	**4**
Spinner B	**1**	0	1	2	3
	2	1	0	1	2
	3	2	1	0	1
	4	3	2	1	0

  b i $\frac{1}{4}$ ii $\frac{1}{8}$ iii 0

4 a

Spinner

1	(1,1)	(2,1)	(3,1)	(4,1)	(5,1)	(6,1)
2	(1,2)	(2,2)	(3,2)	(4,2)	(5,2)	(6,2)
3	(1,3)	(2,3)	(3,3)	(4,3)	(5,3)	(6,3)
	1	2	3	4	5	6

Dice

  b i $\frac{1}{6}$ ii $\frac{1}{9}$ iii $\frac{1}{3}$

5 a

Box B

R	(R,R)	(Bu,R)	(Y,R)	(Ba,R)
Y	(R,Y)	(Bu,Y)	(Y,Y)	(Ba,Y)
Bu	(R,Bu)	(Bu,Bu)	(Y,Bu)	(Ba,Bu)
	R	Bu	Y	Ba

Box A

  b i $\frac{1}{12}$ ii $\frac{1}{4}$ iii $\frac{3}{4}$

6 $\frac{3}{5}$

7 a $\frac{1}{6}$ **b** $\frac{1}{10}$ **c** $\frac{1}{60}$
  d $\frac{1}{20}$ **e** $\frac{1}{12}$

5.2 Get Ready

1 a $\frac{1}{4}$ **b** $\frac{3}{5}$
2 a $\frac{3}{4}$ **b** $\frac{7}{20}$

Exercise 5D

1 a $11\frac{1}{6}$ **b** $55\frac{5}{6}$
2 £320
3 £1740
4 No. She will need to save a little more. Her mother gives her £56. This means she has a total of £276.
   She needs another £4.
5 £187
6 20%
7 £674.92
8 Mr Jackson is right. The spade should cost £19.80.
   The salesman has taken 22% off the price instead of 10% then 12%.
9 Sophie should use the second account as: First gives £848.70 Second gives £848.72 She will get 2p more interest!
10 It is a bargain price. In fact James could have priced it a bit higher.
   After 1 year its value is £5225 after 2 years its value is £4180 and after 3 years its value is £3344.

5.3 Get Ready

1 a 0.4   **b** 0.65   **c** $\frac{3}{5}$ **d** $\frac{1}{6}$

Exercise 5E

1 a $\frac{3}{10}$ **b** $\frac{2}{5}$ **c** $\frac{7}{10}$
2 0.9
3 a 0.5   **b** 0.5   **c** 0.65   **d** 0.7
4 a $\frac{3}{13}$ **b** $\frac{3}{13}$ **c** $\frac{1}{13}$
  d $\frac{6}{13}$ **e** $\frac{4}{13}$ **f** $\frac{4}{13}$
5 a $\frac{1}{2}$ **b** $\frac{7}{18}$ **c** $\frac{2}{9}$ **d** $\frac{2}{3}$
6 a $\frac{1}{26}$ **b** $\frac{1}{2}$ **c** $\frac{5}{52}$
  d $\frac{7}{26}$ **e** $\frac{7}{26}$
7 0.35
8 The events are not mutually exclusive. The probability is $\frac{1}{36}$.

Exercise 5F

1 $\frac{11}{72}$
2 $\frac{4}{5}$
3 0.35
4 0.25
5 a $\frac{1}{13}$ **b** $\frac{12}{13}$
6 a i 0.65   ii 0.85
  b $0.35 + 0.15 + 0.5 = 1$
7 a $\frac{2}{9}$ **b** $\frac{7}{9}$ **c** $\frac{5}{9}$
8 0.6

5.4 Get Ready

1 a $\frac{1}{2}$ **b** $\frac{3}{4}$ **c** $\frac{8}{9}$ **d** $\frac{3}{13}$

Exercise 5G

1 $\frac{9}{20}$
2 $\frac{13}{15}$
3 a $\frac{77}{150}$
  b No, red is almost three times as likely as orange.
4 a Students' own work
  b Drop the drawing pin more times, because the greater the number of trials, the more accurate the estimated probability.
5 $\frac{33}{200}$, because the greater the number of trials, the more accurate the estimated probability.
6 Students' own work

5.5 Get Ready

1 a 50   **b** 18   **c** 30   **d** 40.5

Exercise 5H

1 50
2 10
3 30
4 a 20   **b** 65   **c** 80
5

Colour of peg	Red	White	Yellow
Expected number	25	125	100

6 30
7 No, the draw is random so he could be unlucky.
8 The doctor's estimate is a bit high, as the results from the 240 patients suggests a probability of 0.083.

5.6 Get Ready
1 a $\frac{1}{6}$ b $\frac{2}{7}$ c $\frac{2}{5}$ d $\frac{3}{16}$

Exercise 5I
1 0.32
2 $\frac{1}{2}$
3 0.6
4 a 0.65 b 0.85 c 0.5525
5 a $\frac{1}{4}$ b $\frac{1}{16}$ c $\frac{3}{169}$
  d $\frac{1}{676}$ e $\frac{1}{2704}$
6 a $\frac{3}{44}$ b $\frac{2}{11}$ c $\frac{15}{22}$

5.7 Get Ready
1 a $\frac{3}{4}$ b $\frac{8}{15}$ c $\frac{13}{20}$ d $\frac{55}{63}$

Exercise 5J

1 a

Bag A	Bag B	Outcomes

b i $\frac{12}{35}$ ii $\frac{6}{35}$ iii $\frac{18}{35}$

2 a First pencil Second pencil Outcomes

$\frac{3}{10}$ — HB — $\frac{3}{10}$ — HB HB, HB $\frac{9}{100}$
$\frac{7}{10}$ — not HB HB, not HB $\frac{21}{100}$
$\frac{7}{10}$ — not HB — $\frac{3}{10}$ — HB not HB, HB $\frac{21}{100}$
$\frac{7}{10}$ — not HB not HB, not HB $\frac{49}{100}$

b $\frac{21}{50}$

3 a Ryan Ibrahim Outcomes

$\frac{3}{8}$ — orange — $\frac{2}{5}$ — orange orange, orange $\frac{3}{20}$
$\frac{3}{5}$ — red orange, red $\frac{9}{40}$
$\frac{5}{8}$ — red — $\frac{2}{5}$ — orange red, orange $\frac{1}{4}$
$\frac{3}{5}$ — red red, red $\frac{3}{8}$

b i $\frac{3}{20}$ ii $\frac{19}{40}$

4 a i 0.6 ii 0.3
b

1st lights	2nd lights	Outcomes

0.4 — stopped — 0.7 — stopped stopped, stopped 0.28
0.3 — not stopped stopped, not stopped 0.12
0.6 — not stopped — 0.7 — stopped not stopped, stopped 0.42
0.3 — not stopped not stopped, not stopped 0.18

c i 0.28 ii 0.54 iii 0.82

5 a

Spinner A	Spinner B	Outcomes

0.3 — 6 — 0.45 — 6 6, 6 0.135
0.55 — not 6 6, not 6 0.165
0.7 — not 6 — 0.45 — 6 not 6, 6 0.315
0.55 — not 6 not 6, not 6 0.385

b i 0.385 ii 0.48 iii 0.315

6 $\frac{5}{12}$

7 a $\frac{1}{169}$ b $\frac{9}{16}$ c $\frac{3}{8}$
  d $\frac{51}{1352}$ e $\frac{7}{16}$

8 a HHH HTH HHT HTT
   THH TTH THT TTT
  b i $\frac{1}{8}$ ii $\frac{3}{8}$

9 0.243

10 54

5.8 Get Ready
1 $\frac{3}{9} = \frac{1}{3}$

Exercise 5K
1 b i $\frac{7}{15}$ ii $\frac{8}{15}$ iii $\frac{7}{15}$
2 b i $\frac{5}{33}$ ii $\frac{35}{66}$ iii $\frac{15}{22}$
3 a $\frac{3}{14}$ b $\frac{4}{7}$
4 $\frac{121}{315}$
5 0.74
6 1.325×10^{-1}
7 a $\frac{2}{7}$ b $\frac{2}{7}$
8 a $\frac{1}{221}$ b $\frac{1}{17}$ c $\frac{13}{17}$
9 $\frac{37}{49}$
10 0.7816

5.9 Get Ready
1 $\frac{1}{4}$
2 $\frac{2}{5}$

Answers

Exercise 5L

1 $5:6:3$

2 880 g butter, 400 g sugar, 1400 g flour

3 **a** $8:7:11$ **b** $\frac{11}{26}$

4 11

5 **a** $\frac{5}{12}$ **b** $\frac{1}{4}$

6 Lynsey £400 Isla £300

7 700 kg copper 300 kg zinc

8 **a** £285   **b** £52.25

9 **a** 7 mm
   b It could be wrong since this is an extreme weight and is a long way from 4 kg.

Review exercise

1 **a** **i** $\frac{1}{50}$ **ii** $\frac{17}{25}$ **iii** $\frac{1}{10}$ **iv** $\frac{3}{25}$
   b 2%, 68%, 10%, 12%
   c 0.02, 0.68, 0.1, 0.12
   d There are other ingredients not listed.

2 **a** $\frac{4}{11}$ **b** $\frac{4}{11}$ **c** $\frac{7}{11}$
   d $\frac{7}{11}$ **e** $\frac{8}{11}$ **f** $\frac{3}{11}$

3 **a** $\frac{23}{45}$ **b** $\frac{3}{10}$ **c** $\frac{8}{45}$
   d $\frac{37}{90}$ **e** $\frac{8}{45}$

4 $\frac{7}{18}$

5 £2612.50

6 Maggie £700 John £2100 Mark £2800

7 **a** 16   **b** £5.49

8 **a** $\frac{1}{6}$ **b** $\frac{1}{4}$ **c** $\frac{5}{12}$

9 Account 1 gives £200 a year interest which is a total of £1000 over the 5 years. Account 2 gives a final payout of approx. £5938 which is interest of £938.
   Account 2 gives the best return. Provided Mark does not need the interest paying every year he should use Account 2.

10 **a** 97 or 98   **b** 7 or 8   **c** 30

11 Roll the dice 100 times, recording the results. If the dice is fair, each number should occur approximately 17 times. If Amy wanted to be more confident in her results, she could roll the dice more times.

12 90

13 **a** If Megan bought all the tickets, she is definitely right. Otherwise, she could be right, but there is no way of knowing because the winning tickets are drawn at random.
   b The winning tickets are chosen at random, so she may just have been lucky.

14 **a** **i** $\frac{b}{12}$ **ii** $\frac{(12-b)}{12}$ **b** 3

15 $\frac{3}{4}$

16 0.5525

17 **a** 96
   b The original set of marked rabbits came from throughout the warren. The marked rabbits spread throughout the warren after being returned. The dye did not come off any rabbits.

18 $\frac{5}{144}$

19 **a**

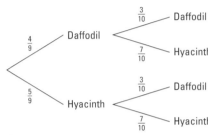

   b $\frac{2}{15}$

20 **a** No. Horse C is more likely to win, because of the four possible outcomes, two are a head and tail, but two heads and two tails are each only one outcome.
   b No. Horse 7 is more likely to win, because of the 36 possible outcomes from rolling two dice, a score of 7 occurs more often than any other score.

21 Number of possible selections of six numbers = $\frac{40 \times 39 \times 38 \times 37 \times 36 \times 35}{6 \times 5 \times 4 \times 3 \times 2 \times 1} \approx 4$ million
   So $P(\text{choosing correct six numbers}) = \frac{1}{4}$ million

22 $\left(\frac{1}{6}\right)^6 = \frac{1}{46\,656} = 2.1 \times 10^{-5}$

23 $\frac{1}{5}$

24 **a** $\frac{1}{12}$ **b** $\frac{5}{18}$ **c** $\frac{13}{18}$

25 $\frac{\pi}{4}$

26 Students' investigations

Averages and range

1 a Add 2   **b** the same   **c** double
  d double
  e 1   2   3   4   5   8   19

2 72

3 The new mean is two larger.

4

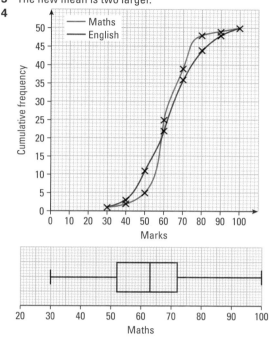

The median for Maths (62) is larger than the Median for English (60).
The Maths marks are more varied as The IQR is 20 compared with the English IQR of 15.

5 a May's average hours of sunshine are higher than June so May had more hours of sunshine than June
The number of hours of sunshine were more variable in June as the range was bigger.

  b   **i** The mean and range take into account all of the data.
     ii The median and IQR are not affected by extreme values.

Intepreting and displaying data

1 Take a random stratified sample.
Stratified by school 9 from Avon, 10 from Moorside, 14 Heaton, 12 Moortop, 15 Brambell
Students must be chosen randomly (names in a hat)
Or
Take a systematic sample $\frac{5720}{60}$ students = 95
Put numbers 1–95 in a hat and draw out a number.
Go systematically through each school numbering all students and selecting the students with numbers divisible by the number drawn.

Or
Put all names in a hat and draw out 60 students – this sample may not contain student from every school.

2

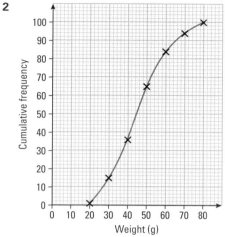

  11 pebbles

3 10 workers

4 75

Probability

1 $\frac{3}{8}$, 0.375 or 37.5%

2 20%, $\frac{1}{5}$ or 0.2

3 0.1, $\frac{1}{10}$ or 10%

4 0.625, $\frac{5}{8}$ or 62.5%

5 $\frac{1}{3}$, 33% or $0.\dot{3}$

6 When Mona has eaten one sweet there are 19 sweets left in the bag. There are still x chocolates. So the probability of Sam eating a chocolate is x divided by 19 (to be formatted correctly in the book).

7 The probability of the first sock picked being x, depends on the total number of socks in the drawer (i.e. the number of grey socks + the number of black socks) and the number of black socks at the start. When one black sock has been removed, the number of black socks and the total number of socks will both be lower by one. Therefore the probability of picking a black sock the second time will not be x.

Snowdon

1 The 9 am train is cheaper by £6. (£46 compared to £52)

2 Rashid: time = 2.75 hours, distance = 7 km, speed = 2.55 km/h
Chelsea: time = 3 hours, distance = 9 km, speed = 3 km/h
Rashid is not the faster walker.

Answers

3 a

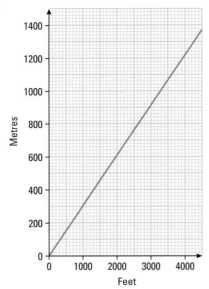

b Height of Snowdon = 3560 × 0.3048 = 1085 m
Ben Nevis is higher.

Music sales

1 Sales in £ millions:

	2006	2007	2008	2009
Downloads	69	115	169	166
CDs	134	135	98	95
Total	203	250	267	261

The total sales between 2006 and 2008 increase each year and then decrease slightly in 2009.
CD sales stay approximately the same from 2006 to 2007 and then decrease, whereas download sales increase significantly each year before steadying out in 2009.

2 Various answers are possible, for example, two frequency polygons.

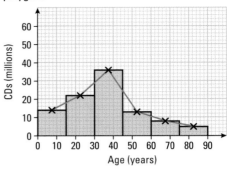

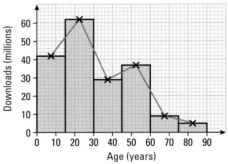

3 Suitable questions for a data collection sheet, for example:
Are you male or female? Male ☐ Female ☐
How old are you?
0–14 ☐ 15–29 ☐ 30–44 ☐ 45–59 ☐ 60–74 ☐
75–89 ☐ 90 or over ☐
How much do you spend on music each month?
£0–4.99 ☐ £5–9.99 ☐ £10–14.99 ☐ £15–19.99 ☐
£20 or more ☐
What kind of music do you buy most often?
Pop ☐ Classical ☐ Jazz ☐ Folk ☐ Other ☐

Index

Index